Controlling Dietary Fiber in Food Products

Controlling Dietary Fiber in Food Products

Leon Prosky
Division of Nutrition, HFF-268
Food and Drug Administration
Washington, D.C.

Jonathan W. DeVries
General Mills, Inc.
James Ford Bell Technical Center
Minneapolis, Minnesota

VNR VAN NOSTRAND REINHOLD
New York

An AVI Book
(AVI is an imprint of Van Nostrand Reinhold)

Library of Congress Catalog Card Number 91-23996
ISBN 0-442-00239-4

Printed in the United States of America

Van Nostrand Reinhold
115 Fifth Avenue
New York, New York 10003

Chapman and Hall
2-6 Boundary Row
London, SE1 8HN, England

Thomas Nelson Australia
102 Dodd Street
South Melbourne 3205
Victoria, Australia

Nelson Canada
1120 Birchmount Road
Scarborough, Ontario M1K 5G4, Canada

16 15 14 13 12 11 10 9 8 7 6 5 4 3 2 1

Library of Congress Cataloging in Publication Data

Prosky, Leon.
Controlling dietary fiber in food products / by Leon Prosky and Jonathan W. DeVries.
p. cm.
Includes bibliographical references and index.
ISBN 0-442-00239-4
1. Food—Fiber content. 2. Fiber in human nutrition.
I. DeVries, Jonathan W. II. Title.
TX553.F53P76 1991
613.2'8—dc20 91-23996
CIP

Leon Prosky would like to dedicate this book to Heather, Melissa, and Rebecca. Jon DeVries would like to dedicate this book to Sharon and Jon Jr., for their sustained understanding and patience during the preparation of this manuscript.

Both authors would like to acknowledge all the efforts of the numerous devoted scientists who have contributed so greatly to the body of knowledge on this subject over the last 25 centuries.

Contents

Notice of Disclaimer

The opinions expressed in this book are solely those of the authors and do not affect or represent regulatory policy of the Food and Drug Administration or General Mills, Incorporated.

Preface

The study of the properties, effects, and levels of dietary fiber in foods has achieved great importance in nutrition and food technology during the past 15 years. Recently, the Congress of the United States enacted legislation that makes labeling dietary fiber in foods compulsory. With this in mind, the authors set out to write a short book detailing 1) the history and properties of food fiber, 2) the evaluation of the current methods used in measuring dietary fiber, and 3) the method of choice in the measurement of dietary fiber. This book also contains some discussion of the marketing of dietary fiber products, including additives. Accompanying the text are a number of tables of food values for dietary fiber obtained by using the AOAC method of analysis in a variety of laboratories in this country and abroad.

1

Properties of Food Fibers/Fibers in Food Products

DEFINITION OF DIETARY FIBER

Superficially, the concept of dietary fiber is easy to grasp. Fiber is a material found in or associated with food, but it is not digested and does not provide energy (calories) or building blocks for the structural growth and maintenance of the consuming organism. Many of the familiar plant-based fibers in our lives, such as cotton, wood, banana peels, corn husks and cobs, wheat hulls, grass, and so on, are very "fibrous" in appearance. Not only are they nondigestible, they are difficult to consume as well. Many of these fibers have important properties unto themselves, which make them very useful for clothing, shelter, and so on, but not as food. If nature was so clear cut that the digestible and nondigestible portions of the plant were also the consumable and the nonconsumable respectively, there would be no reason to concern ourselves with an appropriate definition of dietary fiber. It would be self-evident.

Nature, however, does not provide such a simple differentiation when it comes to the human food supply. Most humans know that portions of the food they consume are nondigestible, that is, that the feces they excrete are made up not only of waste byproducts resulting from the normal conversion of digestible foodstuffs to energy, but also of materials that passed through the alimentary system untouched by digestive enzymes. What is not realized is that, in addition to the materials that are obviously fibrous in appearance, there are food components, such as pectin and β glucans, that do not appear to be fibrous at all, are even water-soluble, but that are not digested by the enzymes of the alimentary system. Therein lies the dilemma of deriving an appropriate definition for these indigestible food components.

As early as the 1800s, the term fiber was used as it related to animal forages. This fiber, termed crude fiber, was the residue left over after extensive and strenuous chemical digestion of the sample. Since this material was generally considered totally nondigestible and therefore of little benefit to humans, food nutrition tables sometimes did not report the crude fiber content of the foods contained therein. When it *was* reported, the main purpose of the assay was to determine the portion of the food or forage totally nondigestible by nearly any means and thereby adjust the calculated energy (calories) content of the food accordingly.

T.L. Cleave (Cleave 1956; Cleave and Campbell 1966) noted a marked contrast in the types of diseases observed in the residents of underdeveloped countries, when compared to those of the residents of industrialized nations. These disease states appeared related to the people's diets. Residents of underdeveloped countries had diets consisting of high levels of whole plant materials and/or other whole foods. In contrast, people in industrialized nations consumed very low levels of whole food or whole plant materials, due to their having convenient access to refined foods. Cleave further noted a relationship between these consumption patterns and a number of diseases that he termed "fibre deficiency syndromes."

Trowell, in 1972 (Trowell 1972a, b, c), wishing to show the inadequacy of crude fiber figures in food nutrition tables and the importance of analyzing all the nondigestible plant cell wall materials in the diet, suggested the term "dietary fibre," a term coined earlier by Hipsley (1953). This preserved the common term of *fiber*, associated with component nondigestibility, but allowed for a definition that would include all the significant dietary components rather than the leftover residue from the harsh crude fiber measurement technique. Trowell says that the "hallmark of all the substances included in the term 'dietary fibre' was that they were not digested at *all* by the alimentary enzymes of man" (Trowell et al. 1985).

In 1979, at the 93rd annual meeting of the Association of Official Analytical Chemists (AOAC), Prosky and Harland, wishing to standardize both the definitions and methodologies for dietary fiber, announced their intention of seeking a definition and a commensurate method for analyzing total dietary fiber that was suitable for collaborative study. By the 95th annual (1981) meeting of that professional group (which specializes in validating methodology, particularly methodology that is to be used in regulatory decision making or enforcement), more than a hundred responses from the scientific community had been received, with regard to the definition. The preferred definition was that of Trowell (Trowell 1974): "Dietary fibre consists of remnants of the plant cells resistant to hydrolysis (digestion) by the alimentary enzymes of man," whose main chemical constituents are hemicelluloses, celluloses, lignin, oligosaccharides, pectins, gums, and waxes.

This definition is very useful for delineating food portions that are unavailable for conversion to energy or are otherwise necessary for structural growth and maintenance, yet impart important physiological effects in their own right. As shown by the input from the scientific community, this definition accurately reflected the current scientific opinions regarding the constituency of dietary fiber. This definition further lends itself to practical methodology, in that a system or systems need only be devised that will quantitatively segregate the digestible from the nondigestible portions of the food, with appropriate corrections of course for minerals, and so forth.

As with any subject of such a broad scope, alternate definitions for dietary fiber have been proposed. Englyst et al. (1987) proposed dropping the use of "nondigestible remnants," in favor of a more restrictive, strictly chemical (as opposed to physiological) definition. Dietary fiber would thus be defined only as nonstarch polysaccharides (NSP). Methodologically, these nonstarch polysaccharides would be measured by instrumental methods, after they are isolated from an enzymatic digest.

Of particular concern to Englyst is a component of certain foods (in particular, processed cereal grains) that is referred to as resistant starch. This is starch that has in some way become modified during processing so that it is no longer digestible with physiological enzymes. Scientists who do not want to include this fraction in the definition of dietary fiber base their argument on observations of ileostomy patients. In these observations, it is noted that not only do the usual fiber components and resistant starch pass through the small intestine intact, but some portion of the regular "nonresistant" starch may also pass through undigested. It is reasoned that since some normal starch passes through the small intestine undigested in these patients, normal starch cannot be differentiated from resistant starch and, therefore, resistant starch should not be included as part of the definition of dietary fiber.

Such an argument, of course, merely begs the question (Asp et al. 1988). Resistant starch can be readily differentiated from regular starch, as evidenced by the fact that the one form (resistant starch) is consistently not digested by physiological enzymes, while the other form (regular or nonresistant starch) is. The fact that a portion of the regular (nonresistant) starch was not digested by a particular patient, on a particular diet, in a particular study does not necessarily mean that all the nonresistant starch will not be digested in an alternative situation, where the food is prepared differently, the makeup of the meal other than the starch source is different, or the health state or the demand for calories in the individual consuming the starch is different. The relationship between the relative quantity of consumed regular starch that is digested and the calorie demand of the individual consuming it has not been established. It can be reasonably assumed that this starch is available for digestion, if the consuming individual's metabolism requires it.

On the other hand, resistant starch is simply not digested enzymatically. Therefore, it is reasonable that this starch would not be available for normal metabolism to occur in the individual consuming it, even if that individual required it. In fact, a recent study is reported to have shown that all the starch in the residue from mixed diets isolated in the enzymatic-gravimetric method, developed collaboratively by the AOAC, escapes digestion and absorption in the small intestine (Schweizer 1989). Only after being dispersed in a special treatment step with a powerful solvent, like dimethyl sulfoxide (DMSO), is the resistant starch susceptible to enzymatic digestion. While this treatment is known to disperse the resistant starch as part of the analytical procedure that was developed to quantitate the level of nonstarch polysaccharides (NSP), the extent of the effects that the DMSO may have on other normally nondigestible portions of the food is not well studied. The DMSO treatment step may in fact lead to erroneous results, when determining either NSP or dietary fiber, if the DMSO in any way affects the enzymatic digestibility of other fibrous components.

Furthermore, using dietary fiber definitions and resulting methods that selectively exclude portions of (or underestimate the proportion of) the food that is nondigestible leads to erroneous estimates of the food's caloric content. This results in misinformation regarding the caloric content of the foods being analyzed, which in turn supplies inaccurate information for nutrition tables and nutrition labels. In cases of obesity, where the caloric content of the food is being monitored for the purpose of weight control, the most accurate information available should be used. Quantitating the entire nondigestible portion (or proportion) of the food is the best way to provide this information.

The study of dietary fiber, its components, and their relationship to various health states and diseases is presently a vigorous, ongoing process. It is now, and will be more so in the future, desirable to investigate the potential impact of individual characterized components of the dietary fiber on human health and nutrition. Rugged and effective analytical methods for characterizing and quantitating these individual fractions will be necessary in order to fully assess these impacts. Methods that can effectively aid the scientific community in reaching a consensus on the cause/effect relationships of dietary fiber in total or as components will be a major asset in understanding these relationships.

Scientific definitions, as with all definitions, are subject to change, of course. But until the cause/effect relationships described earlier are established, the current definition will serve us well in defining and quantitating that portion of the food for which various physiological responses have been observed and reported. For the present, dietary fiber defined as "consisting of the remnants of plant cells resistant to digestion by the alimentary enzymes of humans" is the appropriate definition.

EDIBLE FIBER

The majority of the epidemiological observations, the research studies seeking various physiological responses, and the resulting recommendations to make changes in eating habits based on the effects attributed to dietary fiber have focused on the nondigestible portion of plants. In many cultures, however, plant matter makes up a small portion of the daily food intake (Van Soest, Lewis, and Robertson 1988). Nonetheless, a portion of the diets consumed by members of these cultures is nondigestible. Trowell et al. in 1978 applied the term "edible fiber" to nondigestible fibrous materials found in foods, no matter what their source, implying a nondigestible, fiber-like material that is safe to be eaten. This includes not only the plant-based food components defined above, but also undigested animal polysaccharides, such as connective tissue, and undigested pharmaceutical polysaccharide preparations, such as ispaghula husks (husks from the Plantago ovata seed, commonly known in the United States as Indian Psyllium). Broad-based usage of the term "edible fiber" and its definition could be of significance. When calculating calorie content of a foodstuff, for instance, utilizing the total quantity of all the nondigestible fraction of the foodstuff would give the most accurate indication of calorie content. Certain components of processed foods (for example, Maillard browning products) are nondigestible by the enzymes of the human digestive system and may therefore be noncaloric. These browning products may produce a physiological response similar and/or complimentary to the dietary fiber fraction, as currently defined. In addition to noncaloric effects, future research on these components may bear out the fact that they provide other beneficial physiological effects.

If desired, the current Association of Official Analytical Chemists (AOAC 1990) methodology could easily be modified, to take this into account, merely by not including a protein correction on the dietary fiber residue. In addition, the nondigestible remnants of non-plant-based foods may exhibit many of the desirable properties of dietary fiber, upon further study (Betschart et al. 1990). The scientific community can probably best serve people's health and nutritional needs by vigorously collecting data and knowledge on the various sources of nondigestible food fractions and not focusing only on the plant-based materials. To this end, using and maintaining the definition of edible fiber as "those remnants of the food resistant to digestion by the alimentary enzymes of humans," in parallel with or in place of the definition for dietary fiber, might be in order in the future.

DIETARY FIBER COMPOSITION

Dietary fiber is a complex mixture of a variety of entities that are resistant to digestion by the alimentary enzymes of humans. Because it is a portion of a natural product, its makeup can vary considerably from source to source. No standardized classification or terminology system has been devised or agreed upon, for the purpose of characterizing various compounds or components of dietary fiber in a consistent way. A number of possible schemes could be developed for this purpose. One possibility is to only use chemical names for each of the components that has been completely characterized (for example, 1,4-β glucans (cellulose) or β-1,3 glucans (β glucans). Another is to classify dietary fiber types based on their function in the plant of origin (for example, cell wall or storage components). Alternatively, one could classify them by their behavior or physiological response in the human (for example, hypocholesterolemic, laxative). None of these systems are completely clear cut. For the present, therefore, it is best to define dietary fiber components using the colloquial terminology of those scientists currently involved with research on fiber itself or involved with research on the physiological effects of fibers on the human system.

Currently, total dietary fiber is split into two main components: soluble dietary fiber and insoluble dietary fiber. Insoluble dietary fiber is plant material that is not digestible by appropriately chosen enzymes that mimic the human alimentary system and is not soluble in hot water. Soluble dietary fiber is food material that is not digestible by appropriately chosen enzymes and is soluble in warm or hot water, but is reprecipitated when that water is mixed with four parts of ethyl alcohol. The soluble fiber and the insoluble fiber each have distinct chemical characteristics and physiological effects. In terms of physiological activity, in general, soluble dietary fiber is more effective in reducing hyperlipidemias, while insoluble dietary fiber is better for alimentary system dysfunctions, such as constipation.

Insoluble Dietary Fiber Components

Cellulose—Cellulose is probably the least soluble of all fiber components, being insoluble not only in cold or hot water, but also in hot dilute acids and alkalis as well. Cellulose, the major structural component of plants, is a glucose polymer bonded in the β-1,4 linkage configuration (as compared to starch, which has alpha 1,4 and 1,6 linkages). The 1,4 β linkage allows the cellulose polymer to crystalize in a linear configuration, with a high degree of intermolecular hydrogen bonding, which gives it substantial shear and tensile strength. Because of its chemical makeup, cellulose can be purified for use as a food ingredient.

Hemicellulose—The name hemicellulose might lead one to believe that it is a precursor or breakdown product of cellulose. Actually, the two have relatively little in common chemically. Both are insoluble in hot water, and both are polysaccharides. What distinguishes hemicellulose from cellulose is the fact that hemicellulose can be dissolved in dilute alkali. The main structure of hemicellulose is composed of a number of monosaccharides, primarily xylose (xylan polymer), glucose and mannose (glucomannan polymer), and galactose (galactan polymer). Attached to this main structure are side chains of glucose, arabinose, and glucuronic acid.

Lignin—Lignin is a highly water-insoluble polymeric material, derived in the plant by polymerization of aromatic alcohols—cinnamyl, syringyl, and guaicyl alcohol in particular. When intricately intermingled with the cellulose and hemicellulose of the plant fiber, lignin increases resistance to degradation and subsequent solubilization.

Cutin and plant waxes—These hydrophobic lipid materials are typically found in the plant structure, closely associated with the structural polysaccharrides or on the outer surfaces of the plant. They are usually present in very small quantities.

Soluble Dietary Fiber Components

Gums—Basically, all soluble dietary fibers are gums from a variety of sources. They are typically (with the possible exception of beta glucans and pectin) present in or used at low levels (<0.5%) in food products. Food gums have very unique functionalities that can improve processing and eating characteristics when formulating high-fiber foods.

β Glucans—β glucans are glucose polymers, wherein the individual glucose monomers are linked together with β 1,4 and β 1,3 linkages, making the polymer resistant to digestive hydrolysis. β glucans are found in significant quantities in oats, rye, and barley. The standard for oat bran, established by the American Association of Cereal Chemists Committee on Oat Bran, is that the bran have a minimum β glucan content of 5.5%, on a dry weight basis. Most of the β glucan present is soluble, although a small amount may be insoluble.

Pectins—Pectins are polymeric substances that are based on a polymer of D-galacturonic acid, linked by α 1,4 linkages. The main polymer has side chains that consist of the sugars, galactose, glucose, rhamnose, and arabinose. Pectins are primarily water-soluble, solubility being somewhat dependent on the degree of esterification of the galacturonic acid, as well as the makeup of the constituent side chains. The primary sources of pectin are citrus fruits and apples, although sugar beet pulp also has a high content of this polymer.

Other Gums—There are a large variety of other gums that contribute to the total dietary fiber content of the food supply. They are used not only to add dietary fiber content to the food, but also to improve food product performance in some respect.

Agar, a seaweed mucilage, is a sulfonated (at the galactose moiety) polymer of anhydro galactose, D and L galactose, and xylose. It is stable at high temperatures, under neutral conditions, and is especially useful as a thickening agent for confections and dairy products.

Alginate, extracted from various species of brown seaweed, is readily soluble in cold and hot water as sodium, potassium, or magnesium salts. Alginates are used for stabilizing colloids, particularly in ice cream-type products, where they assure a creamy texture and retard the formation of ice crystals in the product. The gum is made up of a polymer of glucuronic and anhydro mannuronic acids.

Gum arabic is a highly branched molecule exuded from the acacia tree, whose major sugar constituents are arabinose, galactose, glucuronic acid, and rhamnose. It is primarily used when product flavor stability and long shelf life are desired.

Carrageenan, another seaweed mucilage, is typically marketed commercially as its metallic salts. The gum is basically a sulfonated galactose polymer, the types and properties of the different carrageenans being dependent on sugar linkages and the degree of sulfonation. The gum serves as a gelling and emulsifying agent, as well as a viscosity builder in liquid systems.

Modified celluloses are utilized primarily for their functionality. The most popular modified celluloses are carboxymethyl cellulose (as either the free acid or as the sodium salt), methyl and ethyl cellulose, and hydroxypropyl cellulose (produced by reacting alkali cellulose with sodium chloroacetic acid or with the respective chlorides of the cellulose substituents, respectively). The modified cellulose gums are film formers, viscosity builders, and emulsifiers.

Flax seed gum, a polymer of arabinose, glucose, galactose, and xylose, making up approximately 6% of the flax seed, has been suggested as a substitute for gum arabic in some applications. Whole flax seed, as opposed to the gum fraction, has been incorporated into a number of cereal products for its laxative effect.

Gum ghatti (Indian gum) is a plant exudate of complex structure, consisting of arabinose, galactose, glucuronic acid, mannose, and xylose. Although used as a gum arabic substitute, it is more viscous and less adhesive than arabic. Because of variable supply, its use is not extensive.

Guar gum (or guar flour) gives tasteless, odorless, completely soluble solutions in cold or hot water. The plant from which it is cultivated is used in India as livestock feed. It is essentially a long-chained mannose polymer with galactose side chains. It serves as a protective colloid, film forming agent, stabilizer, and thickener, primarily in thick, viscous, liquid, or spreadable foods.

Gum karaya (also known as Indian tragacanth), a white, slightly acidic material with a mild acetic acid odor, is composed of galactose, galacturonic acid, and rhamnose sugar moieties. It serves as a texturizer, stabilizer, thickener, and emulsifier in foods.

Locust bean gum, synonymous with carob flour, swells in cold water, gives a clear solution in hot water, and is unstable when exposed to warm acids. Its sugar constituents are galactose and mannose. The gum, which takes on a legume flavor when boiled in water, serves as a binder, stabilizer, and thickener. It has been used as a substitute for cocoa, chocolate, and coffee.

Psyllium seed gum, a polymer of arabinose, galactose, galacturonic acid, and rhamnose, was well established as a promoter of laxation as early as the 1930s (BeMiller 1973). Consideration is currently being given to its inclusion in cereal products because of the high level of soluble dietary fiber, and the properties associated there with that it contains.

Gum tragacanth (Syrian gum) has a high resistance to acids and is therefore useful as a stabilizer, under acidic conditions. Its structure incorporates arabinose, galacturonic acid, some xylose, and some small amounts of cellulose, protein, and starch. It has an insipid mucilaginous taste and is useful for making emulsions, thickening, and texturizing.

Xanthan gum, usually sold as its potassium salt, is a microbial gum, from the bacterium Xathomonas campestris, that is a partially acetylated and pyruvated structure of glucose, glucuronic acid, and mannose. A strong film former with good heat resistance, the gum gives highly viscous solutions in low concentration, serving as a stabilizer and thickener in food systems.

DIETARY FIBER CONSUMPTION PATTERNS

Preindustrial Fiber Consumption Patterns

For centuries, the bulk of humankind subsisted on a diet that was high in the nondigestible fractions of plant materials now commonly referred to as dietary fiber. This fiber-rich diet was not necessarily a matter of choice to those who consumed it. Rather, no easy and efficient means had been devised to readily remove the nondigestible portion of the plant matrix from the nutrients and calories that were needed to sustain life and provide energy for the daily tasks associated with survival. Despite the fact that the majority of humanity was consuming relatively high-fiber diets at the time, Hippocrates, in the fifth century B.C., based on his observation of those who apparently were eating more refined foods, is believed to have promoted a diet low in refined foods and therefore high in dietary fiber. The less refined foods, properly prepared, were seen as beneficial for humans, with condiments and confections being considered merely for luxury and gratification of the palate (Adams 1939). This recommendation was probably a result of his unique advantage of being able to observe various segments of the society he lived in. In all likelihood, Hippocrates, in practicing medicine, provided his services to those best able to pay for them—the upper classes of the time. Since relatively little refined food (food with its dietary fiber reduced or removed by some means of processing) was available, the people that Hippocrates was treating were those that would have had access to the refined products. Any epidemiological differences between this class and the subservient classes could be most readily observed by a person in his position, and the conditions observed related back to the differences that must have existed in the diets.

Industrialization's Influence on Fiber Consumption

If necessity can be named as the mother of invention, reduced effort and labor in any task must certainly be cited as the father. The task of consuming the necessary nutrients and calories for daily activities appears to be no exception to the rule. As new technologies became available that segregated the readily digestible fractions of the food supply (believed to be the essential nutrients, such as calories for energy from starch, sugars, and fat, along with protein, vitamins, and minerals) from the nondigestible fractions of the foods, such as fiber, the populations having such technologies available rapidly changed their eating habits and consumed only the digestible fractions.

Humankind has always separated and discarded some nondigestible fractions from food as it is found in the forms supplied by nature. Bananas are consumed with their peels removed. Corn husks are removed before the edible kernels of the corn are eaten. After the kernels are consumed, another fiber-rich fraction of the corn, the cob, is discarded. Hair, sinew, and gristle are often removed from meats, usually before cooking. Even the most primitive cultures perform this level of segregating the edible components from the inedible components (or the digestible from the nondigestible) of their food supply.

However, there are cases of foodstuffs where the basic unit of the foodstuff is small (for example, a kernel of wheat after the chaff (hull) is removed) or the nondigestible fibrous component is an intricate part of the food (for example, the pectin present as an integral part of the edible portion of an orange, after the peel is removed). In these cases, further segregating the nondigestible portion from the rest of the basic unit's components (the bran from the rest of the wheat kernel or the pectin from the orange) is very impractical on a large scale basis, unless special or automated techniques can be used to carry it out.

The early Egyptians apparently had devised methods of sieving ground grains, for the purpose of separating the various fractions of the kernels (Storck and Teague, 1952). Likewise, the Romans also sieved their grist, with the flour being divided into three grades (Storck and Teague, 1952). Homer's Odyssey (about 1000 years B.C.) also refers to grain being sifted after grinding (Kozmin, 1921). Pliny mentions flour being milled in Pompeii and indicates various grades of flour, along with the quantity of bran segregated from the ground wheat.

Although these cultures had the technology to remove the bran fraction from the kernel, the process was laboriously carried out with manual human labor. It is likely that the amount of sieved product produced was so small that very few people had access to the debranned flour. The ruling classes and the elite probably enjoyed the improved texture and flavor of breads made from the higher grade flours, while prohibiting members of other classes from consuming products made from these flours, either by direct edict or by economic forces. The Egyptians had overseers present during the entire process of milling and baking in order to assure that none of the product was lost (possibly consumed by the slaves?) (Storck and Teague, 1952). In Pompeii, the mills were worked primarily by female slaves and later by serfs and criminals. The mill workers were sometimes forced to wear large wooden discs around their necks, to prevent them from reaching their mouth with their hands and eating the flour (Kozmin, 1921). The technique of segregating the grist fractions passed out of common use in the West after the fall of the Roman Empire (Storck and Teague, 1952).

An apparatus, driven by a water wheel, that separated the bran fraction from the rest of the wheat kernel, with a minimal amount of human labor, is said to have been designed by Boller in 1502 (Storck and Teague, 1952). Others added various devices that improved the mechanical separation techniques (which came to be called "bolting"), and as early as 1783, Evans in the United States was designing automatic mills, with bolting as an integral part of the process (Kozmin, 1921). As consumer demand for the products of the technology grew, the technology itself became ever more efficient and automated. It is easy to imagine that this technology was quickly applied to other grain species as well as wheat.

One of the few commonly consumed grains wherein this trend toward separating and removing the bran fraction did not occur was oats. Oat fiber contains a large percentage of water-soluble β glucans, which are more intimately intermingled with other components of the seed when compared to other food grains, and is therefore difficult to separate from the rest of the oat seed. As a result, oat products, such as oatmeal and ready-to-eat oat-based cereals, tended to be whole grain products during the time when the evolving grain handling processes were producing the more refined flours from other grain species.

Concurrent with the industrial revolution producing means for the rapid efficient separation of grain components was the progression of mechanical means for processing fruits and vegetables. Mechanical processing of fruits and vegetables, which are traditionally known for their fiber level when they are in the raw state, often produced food products with significantly reduced fiber content, as in the case of apple sauces or apple juices produced from apples, or of orange juices with the pulp removed produced via frozen juice concentrates from oranges.

Removing the fiber from the agricultural commodity as it was being processed into an ingredient generally resulted in ingredients that produced products that had greater consumer appeal than products produced from those commodities that had their fiber intact. Thus, wheat flour, with the bran removed, could produce a very white bread with a soft, easy-to-chew, cohesive texture. Without the bran fiber present, salivary enzymes could work more quickly on the starch of the bread, to release glucose, giving the bread made from refined flour a sweeter taste than its coarser, fiber-filled counterpart, which was made from whole wheat flour. Early pioneers in the field of dietary fiber advocacy, such as Graham, Kellogg, and Post, observed some of the physiological effects that this gradual shift away from fiber in the diet was having and developed recipes, formulas, and products that would restore the levels of dietary fiber.

Current Fiber Consumption Patterns

Nature and/or history often have unique ways of providing scientists with the means of studying various phenomenon that occur. In the case of dietary fiber, industrialization and the commensurate changes in diet noted in the previous section did not occur at the same rate over the entire globe. In the late nineteenth century and throughout the twentieth century, the so-called first world countries (first world in terms of industrial development) were advancing very rapidly in their technologies and undergoing dramatic dietary consumption pattern changes. The new technologies being developed allowed the easy, rapid, and efficient extraction, from the food source, of the energy and nutrient-bearing portions believed essential for good health and nutrition. Components such as fiber, believed unnecessary for daily sustenance, were merely discarded or diverted as sources of energy for animal species that in turn converted these components into alternative readily digestible and easily available energy and nutrient sources. On the other hand, third world countries lacked the resources and technologies to carry out food component separation. As a result, individuals living under those conditions continued to consume dietary fiber in a pattern similar to that of their ancestors and probably typical of the consumption pattern of the majority of the entire world population, prior to the industrial age.

Observing and reporting this marked contrast in fiber consumption patterns between developed and undeveloped countries and the apparent relationship of diseases to various "fiber deficiency syndromes" in countries experiencing the most rapid industrial development was first done by T. L. Cleave (Cleave 1956; Cleave and Campbell 1966). In order to understand the relevance of epidemiological observations relating fiber consumption to health states, it is helpful to look at the consumption patterns for dietary fiber in various cultures and populations. Bright-See and McKeown-Eyssen in 1984 estimated the per capita crude and dietary fiber consumption of the populations of 38 countries. They arrived at their estimate by multiplying the per capita food supply of a population by the amount of dietary fiber in that particular food supply. The per capita food consumption was probably an overestimate, based on the Food and Agricultural Organization of the United Nations (FAO) report of 1977, on data for the per capita food supply of various populations for the years of 1972 to 1974. For each food category in each country, the FAO estimated human food available from food production, food imports, food exports, food handling losses, and nonhuman food uses. This figure was then divided by the country's population. The estimate did not include a factor for in-home spoilage, waste, or consumption by household pets, and, therefore, the overestimate may have occurred. On the other hand, the dietary fiber content of each food item was probably underestimated, since Bright-See and McKeown-Eyssen used the dietary fiber levels that were determined by Paul and Southgate in 1978. The method used to determine fiber levels included in that report, while quite effective for characterizing various monomeric components of the fibers, has since been shown to underestimate the actual level of dietary fiber, primarily due to insufficient corrections for losses of neutral sugars during sample workup.

Bright-See and McKeown-Eyssen found that dietary fiber consumption ranged from a low of 22.1 g/day in Sweden and the Netherlands to a high of 93.6 g/day in Mexico. There are two items of interest in the results of their tabulation of fiber consumption. The first is the conspicuous absence of data for the populations (most notably those located in the African continent) with which comparisons of western developed cultures were initially and are currently often made, with regard to the lack of diseases resulting from fiber deficiency syndromes. Quantitative data on the dietary fiber consumption of the very populations where the observations that linked health states to fiber consumption were first made is lacking. The second item of interest is the substantial differences in the type or source of fiber intake by the various populations. The dietary fiber intake levels tabulated by the Federation of American Societies for Experimental Biology in 1987 also show data from a limited number of countries. The tabulation is not broken down by fiber source.

PHYSIOLOGICAL EFFECTS OF FIBER CONSUMPTION

Over the past few decades, following the observations and hypotheses regarding "fiber deficiency syndromes" by Cleave (Cleave 1956; Cleave and Campbell 1966), numerous researchers have discovered (or shall we say rediscovered) the relationship between dietary fiber consumption and various health and disease states. However, observation of this relationship is certainly not new. Hippocrates, in the fifth century B.C., is reported to have stated that consuming the whole grain (rough) breads and the foods of the common folk was healthier than consuming the more refined foods of the wealthier classes. Post (of Post Toasties™ fame), Graham (famous for the cracker he formulated that still bears his name), and Kellogg (of Kellogg's Corn Flakes™ fame) all felt that increased dietary fiber in the human diet promoted better health and well being. So the observations of the last three decades are not exactly new. However, there has never been a wealth of information available, as has been the case in the past two decades and currently. More information is available on the sharp contrast in diets in different cultures, fiber makeup, and the various diseases affected by dietary fiber intake. This, combined with advances in research methods, statistical techniques, and data storage, as well as access to rapid processing of data by computers, has revealed a much more complete picture than was possible in the past. This information and the subsequent revelations related to health will continue to unfold even faster in the near future, as each new finding raises an exponential number of new questions, even as it provides an answer to the question to which the research was originally addressed.

That there are numerous solidly established relationships between dietary fiber consumption and health and disease states is unquestionably true. Interestingly, though, finding these relationships should not come as a great surprise to the scientific community. One need only consider for a moment the eons during which humankind had essentially consumed a "rough" diet that consisted primarily of whole foods (whole grains, complete vegetables, fruits, and berries (the edible portions, that is), and whole meat (not the meat or the fat in an isolated form away from the rest of the flesh), and probably lean meat at that (since abundant, readily available supplies of feed grains for the purpose of fattening cattle were not available until recent years). It is not unrealistic to expect that substantial benefits might be accrued by putting a portion of the whole food back into the diet, after it has been absent for a short time, relative to the developmental age of humankind.

Adaption is a normal characteristic of all living organisms. Therefore, the human organism subsisting on whole grains, vegetables, and so forth, would certainly have adapted to and become dependent upon most, if not all, of the various components of those foods being consumed. Many of these dependencies have already been established, such as the body's need for vitamin B1, C, and so on. It is likely, however, that there are additional adapted dependencies that remain undiscovered to date, including some aspects of dietary fiber on the human body. Of course, this means that in the future additional relationships will be found to exist between dietary fiber and health or disease states, or, shall we say, additional evidence of past human adaptation to dietary fiber will be found. Unfortunately, even with current advanced data collection, storage, and manipulation, two key areas of information are unavailable to researchers considering evidence of human adaptation. First, statistical data on disease occurrence, coupled with a clear picture of dietary consumption patterns, are not available. At best, we can conjecture on the life-style of our ancestors as hunters and gatherers. Secondly, because of the prevalence of acute diseases and other life-ending factors, the shorter life span of our ancestors did not allow for the development of many of the chronic diseases that currently appear to be related to dietary factors and that develop with the progressing age of the individual. Therefore, the full impact of early human life-style (including eating habits) on health cannot be fully assessed in terms of the diseases of concern in society today.

When studying the relationships of dietary fiber to its physiological effects, one must always be cautious. Finding that a relationship exists does not necessarily establish a cause/effect situation. One of the difficulties with studying a macronutrient, such as dietary fiber, is that another macronutrient or a micronutrient (perhaps a micronutrient we are not even aware of as yet) might be present in an amount proportional to the macronutrient being studied and, in fact, may be the key factor in the relationship.

Nonetheless, because the presence or quantity of the key factor may not be known, the macronutrient being studied may be credited for the effect. Therefore, researchers, in the enthusiasm and excitement that exists over finding a correlation between consumption and effect, need to be careful to not overlook other potential cause/effect relationships that may exist. An example of this is the observation of an apparent relationship between chromium intake and the diabetic state, discussed in the section on diabetes, following. Increasing the levels of chromium intake in the diet significantly reduced the effects of or delayed the onset of diabetes. Another example is the effect of decreased manganese intake on the increased loss of other minerals, such as iron, copper, zinc, and magnesium, in menstruating women (United States Department of Agriculture 1990a). Since many of the sources (whole grains, vegetables, nuts, and seeds) of manganese in the diet are also considered good sources of dietary fiber, the observed increase in the loss of minerals during menstruation related to dietary changes could be mistakenly attributed to decreased dietary fiber consumption. Whether these or other micronutrients are concentrated, diluted, or directly proportional in the dietary fiber isolates used in some of the research studies is not always obvious or even known. So, while there may be an increase in the number of solid relationships that can be established between dietary fiber intake and various health or disease states, some of the currently established relationships may be disproved in the future, as other factors prove to be the significant cause of the effect observed.

Adverse Effects of Insufficient Dietary Fiber Intake

As advancing industrial development in food processing provided a means for humans to consume their daily requirements of calories in order to provide the energy necessary for day-to-day activities, without the need to consume the nondigestible portions of the plant fibers, a number of disease factors became more prevalent in societies equipped with this industrial capability. As will be discussed more fully in later sections, in Western cultures in particular, the decrease in intake of dietary fiber appears to have resulted in increases in appendicitis (based on early epidemiological observations—currently, a matter of controversy, as will be discussed later), colitis, colon cancer, constipation, coronary heart disease, Crohn's disease, diverticulitis, hyperlipidemia, ileitis, irritable bowel syndrome, maturity-onset diabetes, obesity, and varicose veins.

Benefits of Increased Fiber Consumption

Numerous nutritional and disease conditions respond rapidly and immediately to increases in the level of dietary fiber in the diet. Quick responses to increases in dietary fiber intake are observed where relief is obtained from constipation (improved laxation), in response to insoluble dietary fiber intake, and blood glucose and insulin levels change in response to increased soluble fiber intake. Some conditions respond in a relatively short term. Examples of this would be lowered blood serum and liver cholesterol and lipid levels (again primarily associated with increased intake of soluble dietary fiber) and reduced insulin dependency, in part due to increased peripheral insulin sensitivity. Longer term responses to changes in dietary fiber intake are observed for conditions such as obesity, coronary heart disease (reduced risk thereof, resulting from changes occurring in the short term in many of the key risk factors), diverticular disease, and Crohn's disease.

Fiber and Coronary Heart Disease

Coronary heart disease, a condition that manifests itself over an extended period of time in the patient, is not a disease state whereby a direct relationship with the consumption of particular nutrients or food constituents can be established by direct experimentation. Particular diet regimens would need to be maintained for a much too extended period of time, to observe a cause/effect relationship, and the likelihood of any human subjects submitting themselves to such an ordeal is essentially nil. Rather, the relationship between coronary heart disease and diet must be established by an indirect means.

There are, presently, a vast number of compounds or groups of chemically similar compounds that can be readily and fairly accurately quantitated in the human system. Epidemiologically, coronary heart disease can be (and certainly has been) related to various levels of these analytes. The fact that the prevalence of coronary heart disease increases or decreases relative to one or more of these factors means that it is possible to indirectly predict the likely occurrence of heart disease in a given patient or category of patients.

For example, if a high percentage of coronary heart disease patients show elevated levels of cholesterol in their blood serum, it may be that coronary heart disease leads to elevated blood serum cholesterol or, alternatively, that elevated levels of blood serum cholesterol may (but are not necessarily proven to) cause coronary heart disease. Secondly and better yet, from a health research perspective, if a high percentage of individuals who have a high level of blood serum cholesterol, measured at a certain point in life, go on to develop coronary heart disease at some later date, high blood serum cholesterol may be (but is not necessarily) the cause of coronary heart disease. Thirdly, elevated blood serum cholesterol may not be the cause or effect of coronary heart disease at all, but, in fact, blood serum cholesterol levels may shift only in relation to some other health state in the individual, which in turn relates to either current or future coronary heart disease. In this case, the elevated blood serum cholesterol level can serve as a diagnostic tool for the current disease state or as a predictive tool for a potential future disease state.

In fact, a relationship between coronary heart disease and elevated blood serum cholesterol exists, as do a number of other indirect relationships. Being able to accurately quantitate an analyte, such as cholesterol, can be of significant value when estimating the effectiveness of a diet regimen, as it relates to coronary heart disease. Indeed, if the second or third cases discussed in the preceding paragraph are true, the level of that directly quantifiable but indirectly related health factor can be tracked in concert with the consumption of the diet regimen. If the measured analyte's level changes along with changes in the diet, this is a good indication that the health state that ultimately leads to coronary heart disease is also changing. If a combination of measurable analytes, each of which individually relates to coronary heart disease in some way (albeit indirectly), can all be tracked and found to change with a diet change, the chances of that particular diet change having a significant impact on the health state of the individual on the diet are much greater. If all known indicator factors can be shifted in a positive direction by changes in diet composition, a significant decrease in coronary heart disease ought to be obtainable.

Studies on these effects might appear to be quite straightforward; however, one needs to realize, particularly when looking at disease statistics, that factors other than diet can also be affecting the prevalence and occurrence of the disease. In recent decades, while researchers on the one hand have been investigating the benefits of and making recommendations regarding dietary changes, other researchers have been successfully developing medications and clinical treatment regimens, to reduce the occurrence of and deaths resulting from those same diseases. While this makes the data more complex to interpret, indications of the effects of particular dietary factor can still be extracted from the data. In the case of dietary fiber, consumption level has an impact on a number of factors that relate to coronary heart disease. Increased dietary fiber consumption has been shown to shift all of these factors in a positive (that is, chance of less incidence of heart disease) direction. The measurable factors currently regarded as the most relevant to assessing potential for coronary heart disease that also relate to dietary fiber consumption are as follows.

Apolipoproteins—Apolipoproteins are the proteins that form lipid protein complexes or lipoproteins. Although the word apolipoprotein is not yet common terminology among food scientists and technologists, measuring the levels of particular apolipoproteins in the blood serum has been suggested as a better means to measure the risk of heart disease than measuring low density lipoprotein associated cholesterol (LDL cholesterol, see following) (Vega et al. 1982). In particular, the majority of the protein in LDL is apolipoprotein B (APLP B), and the majority of the protein in high density lipoprotein (HDL) is apolipoprotein A (APLP A)

(Schneeman and Lefevre 1986). Dietary changes leading to decreases in APLP B and increases in APLP A would therefore be desirable as an indicator of reduced heart disease risk. To the extent that APLP A and APLP B may better relate to heart disease or that their quantitation is more accurate and precise than quantitation of LDL and HDL cholesterol, these factors may be useful.

Diabetes mellitus—Patients suffering from diabetes mellitus have a higher risk of coronary heart disease than those that do not (Anderson et al. 1990).

Cholesterol—Elevated levels of blood serum cholesterol (hypercholesterolemia) have been shown to increase the risk of coronary heart disease substantially, with levels of cholesterol greater than 245 mg/dL increasing the risk for men by up to six times over those men with levels less than 180 mg/dL. In the case of women with serum cholesterol levels greater than 260 mg/dL, the risk is increased by up to three times that for women with levels of less than 200 mg/dL. For men, elevated serum cholesterol levels represent one of the highest measurable risk factors (if not the highest) for coronary heart disease. Anderson et al. (1990) presents an excellent discussion on the strength of the relationship between elevated cholesterol levels and increased risk of heart disease, as well as the relative levels of those increased risks.

HDL cholesterol (Cholesterol associated with high density lipoprotein)—Currently, it is generally accepted that an increased HDL cholesterol level decreases coronary heart disease risk. Interestingly, this is the inverse of total blood serum cholesterol, that is, increases in overall cholesterol result in increased risk of disease, while increases in the cholesterol that is associated with high density lipoproteins actually show a protective effect.

LDL cholesterol (Cholesterol associated with low density lipoprotein)—Increases in cholesterol level that are associated with the low density lipoproteins in the blood correlate with increased incidence of coronary heart disease (Eder and Gidez 1982). The LDL component of cholesterol tends to shift in concert with total blood serum cholesterol, a fact that is not too surprising when one considers that LDLs are roughly 1/2 cholesterol and 1/4 protein, as opposed to HDLs, which are roughly 1/2 protein and 1/5 cholesterol. Therefore, LDL cholesterol should correlate similarly to total cholesterol, although in some cases one might expect better correlations to heart disease risk by one factor or the other. Such differences may be due to statistical considerations or possibly to differences in accuracy and precision in analyzing the risk factors for a particular study.

Fibrinogen—High levels of blood serum fibrinogen (a blood clotting factor) increase the risk of cardiovascular disease (Anderson et al. 1990; Kannel et al. 1987; Wilhelmsen et al. 1984).

Hypertension—Coronary heart disease can be reasonably predicted to occur in individuals with high blood pressure (hypertension), with the accuracy of the prediction improving with increasing pressure level (Anderson et al. 1990; Kannel 1986; Kannel, Dawber, and McGee 1980; Kannel and Sytkowski 1987).

Obesity—Individuals who have a significantly higher (130%) body weight to height ratio than the population at large have a much higher risk of developing coronary heart disease (Hubert et al. 1983). Excessive weight to height ratios (particularly when the excess weight is located in the upper body and around the vital organs) are detrimental to the well being of both males and females, with regard to heart disease. In addition to the readily apparent extra strain that the excess weight places on the coronary system, obese individuals often have high levels of LDL cholesterol, low levels of HDL cholesterol, and increased incidences of hypertension and glucose intolerance (Gordon et al. 1977; Kannel and Gordon 1979), all of which are indicators of an increased risk of coronary heart disease in their own right.

Triglycerides—Elevated serum triglycerides may be a risk factor or serve as a predictor of coronary heart disease, particularly in women (Aberg et al. 1985; Gordon et al. 1977), although this is not a universally held opinion (Anderson et al. 1990).

Each of the risk factors for coronary heart disease listed is affected in some way by changes in the amount of dietary fiber consumed. Nearly all credible research studies on each of the measurable risk factors have found that increasing dietary fiber consumption either has a positive impact on that risk factor (that is, lowering the risk of coronary heart disease) or else the change in intake exhibits no effect. Because dietary fiber intake has an impact on each of these risk factors, as well as on other physiological functions, each of the risk factors will be discussed separately in the following sections of this chapter. It should be noted in passing that not only is dietary fiber important in managing the measurable risk factors of coronary heart disease, but recent data suggest that dietary fiber may independently protect against the disease (Anderson et al. 1990).

When considering a disease state as serious as coronary heart disease (the leading cause of death in the United States) and the tremendous financial burden that it imposes on society, it is important to keep two points in mind. The first point relates to perspective. It is important to be thankful that the numerous, horrible, and dread diseases of the past, such as plagues, small pox, polio, scarlet fever, diphtheria, tetanus, scurvy, beriberi, and others, have been rendered innocuous or nearly so by researching, developing, and producing powerful drugs and effective immunizations, by applying efficient and effective sanitation programs to the environment, food, and water, and by properly processing and fortifying our foods. Because of the effectiveness in these areas, the general population grows to an age at which heart disease becomes a major factor. It is hard to imagine being thankful for heart disease, but it is easy to appreciate the opportunity to live to be an age where one must be concerned about the risk of it.

The second point to keep in mind with heart disease relates to risk reduction. Although heart disease may never be displaced from the top of the list of overall causes of death for individuals of all ages (barring, of course, some dread epidemic for which the research, development, and production sector cannot find an antidote in a timely manner), the risk of death or illness from heart disease at a given age can certainly be reduced. Other than genetic factors, the other major factors in heart disease can be affected by the individual involved. In addition to self-managing diet to affect the factors listed earlier, the individual can manage the factors of smoking (or rather, cessation thereof), get adequate physical activity, and appropriately manage the stress in his or her life. As discussed earlier, self-management of diet regimens, including increased consumption of dietary fiber, can be carried out to affect the level (quantity) of these risk factors in a positive manner, which will reduce the risks of coronary heart disease. Many of these factors, such as serum cholesterol, can also be managed by using drugs specifically designed, produced, and marketed for this purpose. On the other hand, increased intake of dietary fiber and other complex carbohydrates provides a pleasant, less costly (Kinosian and Eisenberg 1988), and in many cases less offensive means of reducing the risk of the disease. Simple changes in eating habits, preferably at an early age when such changes might have an impact even on factors such as familial hyperlipidemia, do not need to be a difficult undertaking, especially when one considers the potential benefits of reducing coronary heart disease risk. The food technologist can play a major role in helping the consumer reduce these risks, through efforts in researching, developing, and producing high-fiber food products that have a palatability equal to or greater than their lower fiber counterparts.

Fiber and Diabetes

In 1981, Trowell claimed to have first observed a relationship between dietary intake and diabetes as early as the 1930s, when he observed that the stout Akikuyu nursemaid of a British employer (who provided her with a customary diet of cheap white rice) had an obvious case of non-insulin-dependent diabetes mellitus (NIDDM), whereas none of the rest of his patients (primarily male Akikuyus, presumably on a more traditional high-fiber diet) had developed this syndrome. In a 1960 listing of some 39 diseases (termed "diseases of civilization"), which appear related to advancing industrial development, Trowell (1960) included diabetes, concluding that high-starch carbohydrate and low-fat diets are protective in diabetes. Dietary fiber per se was not mentioned in the relationship. It wasn't until 1973 and 1974 that he postulated a direct relationship between fiber-depleted food and the disease. In particular, Trowell reported a dramatic decrease (55%) in diabetes death rates in civilian males in England and Wales, for the 12 years following 1942. Because World War II and the subsequent years required maximum utilization of available food resources, during these 12 years only a high-fiber (85% extraction rate) "National" flour could be used.

Subsequent to Trowell's postulation of the relationship of dietary fiber to diabetes, Kiehm et al. (1976), Anderson and Ward (1979), Simpson et al. (1981), and Rivellese et al. (1980) designed and carried out a number of studies that showed beneficial effects of high-fiber diets for individuals afflicted with the disease. Numerous beneficial effects were exhibited in both insulin-dependent (type 1) and non-insulin-dependent (type 2) diabetes, by diet regimens restricted in fat and incorporating high levels of, or supplemented with, dietary fiber and complex digestible carbohydrates. The benefits include reduced insulin requirements, increased peripheral tissue insulin sensitivity, decreased serum cholesterol and serum triglycerides, better weight control, and potentially lower blood pressure (Anderson 1987).

In particular, the soluble dietary fibers (either as part of a food or as a supplement well mixed with food) and/or low glycemic index foods appear to exhibit the greatest therapeutic effect for individuals suffering from this malady (Wolever 1990, Jenkins et al. 1990). Glycemic index is defined as simply 100 times the area under the blood glucose concentration curve for the two hours following consumption of the food being tested, divided by the area of the blood glucose concentration curve for the two hours following consumption of 50 g of white bread. In other words, the test subject consumes 50 g of white bread, and blood glucose is measured at regular intervals, usually 30 minutes, for two hours. A plot is made of the blood glucose concentration versus time and the area under the curve measured. The process is then repeated, using 50 g of the food being tested. The area of the test food curve is divided by the area of the white bread curve, and the result is multiplied by 100, to get the glycemic index of the test food. A low glycemic index is preferred because, by definition, the available sugar(s) from the food source being tested are entering the test subject's blood stream at a much slower, smoother pace than is happening with a high glycemic index food. This will result in a decreased short-term insulin demand, a definite benefit to the diabetic. Jenkins et al. (1981) gives an excellent tabulation of the glycemic indices of a wide variety of foods. As discussed above, foods high in soluble dietary fiber, such as legumes and apples, have the lowest glycemic indices and exhibit the greatest therapeutic effect.

Optimistic and substantial research findings such as this, while being good reason for elation at having yet another effective means with which to address yet another disease state, needs to be tempered, lest over-enthusiasm causes oversight of other possible explanations for a particular phenomenon. Recent research by Anderson and Polansky (United States Department of Agriculture 1990b), at the Vitamin and Minerals Laboratory of the Beltsville Human Nutrition Research Center, indicates that increasing the daily dietary intake of chromium to its maximum suggested level is significant in improving glucose tolerance in individuals who have slightly elevated glucose and insulin levels. On the other hand, in individuals with desirable glucose and insulin levels, no change was observed when chromium intake was increased. Whole grains and whole grain (or nearly whole grain) products, such as the high extraction flours postulated by Trowell to be responsible for decreased incidences of diabetes due to their dietary fiber content, would also contain higher levels of chromium than lower extraction flours. In other case studies that related diabetic states to dietary fiber consumption, it is also possible that higher levels of chromium may have been coincidental with the increased dietary fiber. Further investigation of this trace mineral, its relationship to the diabetic state, and the relationship of chromium levels to dietary fiber levels in some of the foods used for research studies will be necessary in order to fully clarify the impact of both the chromium and the dietary fiber on diabetes.

Fiber and Hypertension

Various studies on populations have indicated that vegetarian groups have a relatively lower blood pressure, when compared to those who subsist on an omnivorous diet. Further comparative studies, either comparing vegetarians to control groups or comparing individuals on high-fiber diets with individuals on more traditional diets, have confirmed that increased dietary fiber intake can reduce hypertension. Anderson et al. (1990) have tabulated the results of 16 studies measuring the effect of dietary fiber on hypertension. Fifteen of these studies showed a significant reduction in blood pressure, with the diet change. Since blood pressure correlates with other coronary heart disease factors, such as serum cholesterol (Castelli and Anderson 1986) and obesity, the effect of dietary fiber, while significant, may only be indirect.

Dietary fiber may play only a part in the effect observed on hypertension. Shils (1988) points out that increasing magnesium intake can significantly reduce hypertension. Studies by McCarron (McCarron and Morris 1987; Luft et al. 1989) and Lyle et al. (1987) also indicate a possible mineral intervention in controlling blood pressure. Increasing the level of calcium either in the diet or by supplementation significantly reduced the blood pressure in roughly one half of the hypertensive patients studied. Dietary fiber sources, many of which are whole grains, vegetables, fruits, or legumes, are also very often sources of both magnesium and calcium. On the other hand, there is some concern that high levels of dietary fiber intake may actually result in a negative mineral balance for the consuming individual. Further study will be required to elucidate potential independent, synergistic, or counterproductive effects on hypertension, derived from the fiber components and the mineral components of the foods consumed during hypertension studies.

Fiber and Obesity

If there is one area of technology in which humankind has made spectacular strides, it is in food production. This spectacular advance has resulted in the production of enough food to supply the calorie needs of the whole human population and then some. Political situations may result in the scarcity or nearly total lack of food for some populations, but the capabilities for production are there. Not only is the technology available to produce food in its raw or natural state, but the technology also exists to process the food into various forms that are not only just palatable, but extremely appealing. It might not strike us, then, as being unusual that as the availability of palatable, appealing foodstuffs increases, an ever-increasing segment of the population begins to suffer from an excessive intake of calories, relative to bodily energy demand. The resulting gains in body fat eventually lead to obesity.

To be sure, within any natural system, the weight distribution, as well as the weight to height ratio distribution, will essentially fit a Gaussian curve. (The curve was, after all, derived from observing occurrences of natural phenomenon.) Therefore, statistically, there will always be a number of relatively overweight individuals, just as there will be relatively underweight members of the population. Of concern, however, is the fact that in recent years there appears to have been a substantial shift in the weight to height ratio curve. While the curve itself may appear to be normal in distribution, there is a definite shift toward an increase in the ratio of weight to height. In many cases, this means that individuals at the top end of the distribution are reaching weights that can have a significant negative impact on their state of health.

The optimal weight for a particular individual might not fall exactly in line with the weight to height ratio tables supplied from a number of sources; therefore, judging what constitutes obesity must be done with caution. In an ideal situation, all health risk factors, such as coronary heart disease, diabetes, hypertension, stroke, and cancers, that are impacted by excess weight could be properly assessed and quantitated, with regard to the relative impact that excess weight imparts to that particular risk. The quantified risk response factors could then be summed, to provide a truly quantitative indicator of what ratio of weight to height constitutes obesity. Unfortunately, such a specific assessment is not yet attainable. Nevertheless, a look at some of the available information relating weight to various disease states provides guidelines that can be followed in order to reduce some of the known risk factors.

Simopoulos and Van Itallie (1984) concluded that overweight individuals die at a younger age than those of average weight. In particular, individuals who were overweight during their younger years appear to have a substantially higher risk of reduced life span. In 1985, a health consensus development conference of the National Institutes of Health (National Institutes of Health 1985) concluded that sufficient evidence existed of cause/effect relationships between obesity and an increased risk of morbidity and mortality, for treatment for this health state to be in order. In this case, obesity is defined by implication. That is, obesity should be treated in certain circumstances. These circumstances include the presence of 20% or greater excess body weight (apparently as compared to the general population) and a family history of existing risk factors of maturity onset diabetes, hypertension, hyperlipidemia, cardiovascular disease, gout, functional impairment, or a history of childhood obesity.

The 1988 Surgeon General's report on Nutrition and Health (U.S. Department of Health and Human Services 1988) arrives at a similar conclusion. Obesity is highly prevalent in the United States, with 25% of the adult population overweight; of the overweight individuals, 40% are severely obese. The fact that this excess weight plays a role as a risk factor in diabetes, hypertension, coronary artery disease, stroke, gallbladder disease, and some types of cancer suggests that reducing the average weight of the general population would improve the overall health state of the population of this country.

It has been estimated that 25% of American children are overweight (Forbes 1975). The immediate health implications to the young are probably not significant, but the psychological and social burdens are (Brownell 1984). Overweight children are viewed as failures at self-control and acquire a variety of unflattering labels that tend to destroy their self-image and self-confidence. Worse yet, some 80% of obese children are likely to become obese adults (Abraham and Nordsieck 1960), with all the consequent health complications that are likely to develop.

Overall weight control and a reduced number of individuals who become obese, either as children or as adults, is a worthwhile objective, based on the expected reduction in the disease states for which obesity is a risk factor (American Dietetic Association 1989). The cost of clinically treating these disease states is tremendous, both to the individual and to society. Reducing the levels of calories, particularly calories from fat as part of a well balanced diet, should be an important part of a regimen for weight reduction that also includes proper exercise. The inverse relationship between fiber intake and fat intake (and subsequent reduction in calorie intake) means that high-fiber foods can play an important role in a weight reduction or weight control program. The British Nutrition Foundation (1990) has tabulated the results of 14 studies that investigated the effect on weight loss of a variety of dietary fiber sources included in the diet. Of the 14 studies (conducted by 11 different researchers on some 344 subjects), 10 (over 70%) showed increased (although modest) weight loss as a result of the effects of increased dietary fiber consumption.

Fiber may exhibit its greatest effect in weight control by promoting a sensation of satiety while consuming the fiber-containing food (Holt et al. 1979). The additional noncaloric bulk in the food makes eating more enjoyable and satisfying, and because the fiber also increases fecal bulking, can reduce the incidence of uncomfortable constipation that often accompanies low-calorie diets.

Fiber and Satiety

One of the major actions of dietary fiber is on the satiety of the individual. If one takes a pragmatic view of many of the effects of increased dietary fiber in the human system, this one factor alone may contribute significantly to the improved health state that results. To the extent that satiety decreases the excess intake of calories and the excess intake of other components of the diet, one would expect to see a decrease in the disease states that relate to this excess consumption. The action of fiber on mechanisms expressing hunger and satiety are of major importance. Most reputable weight loss regimens recommend that the individual wanting to lose weight should eat slowly, to give the food time to enter the system and have an effect. Dietary fiber increases the amount of time and effort necessary to chew the food before swallowing; therefore, the food consumption rate is slowed. As a result, mealtime will be increased, and the individual will feel satisfied while consuming less calories than if he or she had been able to eat more quickly. After ingesting a meal, the amount of time spent by the high-fiber food (particularly those high in soluble dietary fiber) in the stomach is increased before gastric digestion is complete, resulting in a prolonged feeling of satiety (Story 1980; Kritchevsky 1988). In addition, pectins and certain gums affect the transit time of food being digested in the small intestine and should produce a similar feeling.

In a study conducted by Burley, Leeds, and Blundell (1987), females of normal weight consumed breakfast meals of high (12 g) and low (3 g) dietary fiber on two separate occasions. Measurements were made on fullness, hunger, desire to eat, and prospective consumption. Participants were offered a lunch food two and one half hours after each breakfast and requested to eat as much as they liked. The breakfast meals consisted of cereal and toast, representing an ordinary breakfast in England and a typical breakfast for the women in the study. In the high-fiber breakfast, bread fortified with guar gum was supplemented with Kellogg's Bran Flakes™. The low-fiber breakfast consisted of white bread and Kellogg's Corn Flakes™. Despite the lower energy content (90 kilocalories) of the high-fiber breakfast, there were no significant differences in any of the measurements taken, with the exception of the fullness ratings, which were higher after the high-fiber meal. At lunchtime, the intake of the ad libitum food was similar after both breakfasts, despite the reduced calorie content of the high-fiber breakfast. One interpretation of the study is that the fiber (guar gum plus wheat bran) intensified the satiating power of the reduced level of calories available in the high-fiber breakfast. That is to say, the increased fiber content of the meal was making up for the loss of satiating power normally associated with the 90 kilocalorie deficit. On the other hand, Levine and coworkers (Levine et al. 1989), in two studies, found that eating two ounces of a high-fiber cereal for breakfast, followed by an ad libitum buffet lunch, resulted in 12.6% and 17.9% reductions, respectively, in total calorie intake, when compared to a breakfast consisting of two ounces of a low-fiber cereal.

In a second study by Burley, Blundell, and Leeds (1987), females of normal weight to height, who had fasted overnight, were fed equicaloric high (30 g) and low (3.3 g) fiber lunches that had similar macronutrient content but vastly different amounts of dietary fiber. The different fiber levels showed no significant effect on hunger, desire to eat, fullness, or prospective consumption. Both the low-and high-fiber lunches were quite substantial lunches and therefore had very potent satiating effects. Following the lunch, the volunteers recorded all food and drink consumed between then and bedtime. Analysis of those records revealed that some 590 fewer kilocalories per volunteer were consumed following the high-fiber lunch, when compared to the low-fiber lunch—a definite intensification of satiety as a result of the high-fiber. Whether the effect demonstrated here is linear (that is, will a lunch with 16.85 g of dietary fiber reduce subsequent calorie consumption by 295 kilocalories?), or if there is some minimal fiber consumption point at which the effect is observed, has yet to be determined.

The whole question of optimizing the dietary fiber delivery system to achieve satiety is open to further study, with almost endless possibilities. The dietary fiber can come in the form of naturally high-fiber foods (such as cereals, fruits, and pulses), in the form of products that have been supplemented with fiber (such as breads with the wheat flour replaced in whole or in part by barley or oat flours), or in the form of a diet supplement (such as fiber isolates in powder, capsule, or tablet form). An excellent chapter describing the issues of satiety and the details of current thinking on the satiation process has been written by Burley and Blundell (1990).

Fiber and Serum Fibrinogen

As indicated in the discussion on coronary heart disease, a positive correlation exists between elevated serum fibrinogen (a blood clotting factor) and the risk of coronary heart disease. Several studies reported in the last ten years indicate that increased dietary fiber intake results in reduced levels of blood serum fibrinogen (Fehily et al. 1982; Fehily et al. 1986; Koepp and Hegewisch 1981) and that plasma-clotting factors can be reduced with a high-fiber reduced-fat diet (Simpson et al. 1982). However, there is disagreement as to whether the effect observed is due to the fiber itself or to other compounds often associated with the fiber, such as omega-3 fatty acids (Spiller 1990). Effects of consuming relatively high levels of vitamin K (a proclotting factor) in concert with increased fiber consumption have not yet been studied. This situation could conceivably arise if certain vegetables high in vitamin K are ingested as the fiber source, as compared to the cereal fiber sources used in the relationship studies.

Fiber and Serum Hyperlipidemia

Serum hyperlipidemia actually refers to the elevated level of a number of entities in the blood serum that relate back in some way to an increased risk of coronary heart disease. This term includes elevated cholesterol level (hypercholesterolemia), elevated levels of serum triglycerides (hypertriglyceridemia), and elevated levels of cholesterol associated with low density lipoprotein and the corresponding apolipoprotein B. In the category of lipids that affect coronary heart disease, one lipid that exhibits an effect counter to the others is cholesterol associated with high density lipoproteins. This HDL cholesterol and its associated apolipoprotein A are actually inversely related to coronary heart disease risk; therefore, hypolipidemic levels of HDL and APLP A are of concern.

Fiber and hypercholesterolemia—Soluble dietary fiber supplementation of the diet can significantly reduce serum hypercholesterolemia (Bell et al. 1990, Federation of American Societies of Experimental Biology 1987). Anderson et al. (1990) has tabulated studies completed on oat bran, oat meal, rolled oats, guar, xanthan gum, locust bean gum, karaya gum, gum acacia, pectin, psyllium, soy polysaccharides, beans, fruits, and vegetables that, overall, show hypocholesterolemic effects that are produced by the increased consumption of these fiber sources. Concentrated sources of dietary fiber, particularly the soluble fraction, can reduce total serum cholesterol by approximately 15% in the short term and by 20% to 25% in the long term, when the diet is supplemented with them. Oat products in particular (rich in water-soluble, nondigestible, β glucans) have more potent cholesterol-lowering action than other food products (Anderson and Siesel 1990). Oat bran therapy is considered to be more cost effective than the traditional hypercholesteremic treatment drugs cholestyramine and colestipol, in treating hypercholesteremic individuals (Kinosian and Eisenberg 1988). Insoluble dietary fiber sources, such as wheat bran, are not generally considered to have a significant hypocholesterolemic effect, although studies do indicate some serum cholesterol reduction (Anderson and Siesel 1990). Dietary fiber fermenting in the large intestine releases short chain fatty acids—acetic, propionic, and butyric acids, in particular. These acids in turn appear to inhibit the synthesis of cholesterol and increase the clearance of cholesterol associated with low density lipoproteins (LDL cholesterol) (Chen and Anderson 1986). This may be a mechanistic explanation for the effects of dietary fiber in reducing serum cholesterol.

Fiber and LDL cholesterol—In nearly all the studies discussed in the previous section that were carried out to determine the hypocholesterolemic effects of various fiber sources, the LDL cholesterol, when it was quantitated and tracked, followed the same trends of the blood serum cholesterol. In other words, a dietary fiber source that reduced total cholesterol to a significant extent also reduced LDL cholesterol to a significant extent as well. As discussed above, LDL cholesterol may be reduced in the blood serum, due to the action of short chain fatty acids, resulting from the dietary fiber fermenting in the intestines.

Fiber and HDL Cholesterol: LDL/HDL Ratio—In the studies where HDL cholesterol was measured, its level remained almost constant, with only slight increases or decreases occurring as a result of an increased intake of soluble dietary fiber. Perhaps more important than the changes noted for the actual level of HDL cholesterol is the fact that the ratio of LDL to HDL cholesterol typically decreased, a factor also considered important in reducing the risk of coronary heart disease. In this regard, oat products and, in particular, oat meal appears to have the greatest impact.

Fiber and Serum Triglycerides—The results regarding the desired lowering of serum triglycerides by increasing the intake of either soluble or insoluble dietary fiber have been mixed, some studies showing no change, some showing an increase in triglyceride levels, and some showing a decrease. Further studies are needed to further elucidate the effects of dietary fiber on serum triglyceride levels. If dietary fiber ingredients were produced that would effectively displace a significant amount of saturated fat in the diet, perhaps short- and long-term studies could be carried out in order to better assess this factor.

Fiber and the Alimentary Processes

Walker and Burkitt (1985) point out that, until recently, a good deal of the time spent by practioners of the art of healing was spent dealing with disorders of the alimentary system and, in particular, the disorders of bowel movement, like constipation and purging. Adams (1939) points out that Hippocrates believed that defecation should occur two or three times daily, to maintain good health. Numerous others since that time have noted the positive effects of normal bowel function on the individual, as to their physical and psychological health. It is this area, perhaps less glamourous relative to some of the more intricate research involved with reducing the risk of coronary heart disease, for example, that stood out to those observing changes in human health related to dietary changes. Abnormal function not only manifests itself in many ways, but nearly all of the manifestations are affected by the level of fiber intake in the diet, as will be discussed in the following sections.

Fiber and Appendicitis

Appendicitis is a malady that has strong epidemiological evidence to support a positive casual relationship with the consumption of dietary fiber. However, Walker and Burkitt (1985) conclude that, while a plausible link can be envisioned, a step-by-step mechanism for occurrence has not been worked out. Nor have controlled studies yet confirmed the epidemiological evidence, although changes in the makeup of the fecal material as a result of dietary fiber intake changes may be indicative of an effect. Further, Walker and Burkitt point out that relationships have been suggested between appendicitis and other "aetiological factors," but the extent of the relationships "precipitating factors is quite unknown."

Fiber and Colitis

See irritable bowel syndrome.

Fiber and Colon Cancer

The relationship between dietary fiber and colon cancer has been the primary focus of studies seeking a relationship between dietary fiber intake and cancer. Colon cancer [the incidence and mortality rate has not changed significantly for the past 50 years in technologically advanced countries and is second only to lung cancer (Lanza, 1990)] is a disease in which one would anticipate the greatest impact of dietary fiber to occur, since dietary fiber has a dramatic impact on other functions of this organ. Burkitt, in 1971, postulated that large bowel cancer could be prevented, or the incidence of cases reduced, by increasing the level of dietary fiber consumed by individuals in a population. There appears to be a number of factors involved with or relating to large bowel cancer incidence or its reduction, as outlined by Cummings (1985). The factors are: stool weight being increased by dietary fiber, colonic contents diluted by dietary fiber, intestinal transit time, influence of fiber on colonic microbial metabolism, fiber-induced effects on fecal enzymes, influence of fiber on fermentation in the colon, effect of fiber on production/distribution pattern of short chain fatty acids produced in the colon, fiber's effect on colonic pH, increases in fecal nitrogen excretion brought about by increased dietary fiber intake, and dietary fiber's impact on bile acids and mutagens present in the colon. Fermentation in the large bowel, which produces short chain fatty acids, actually affects many of the other factors (for example, stool weight and subsequent dilution by osmotic effects and acidification, leading to lower colonic pH (Dreher 1987).

Indications that the risk of colon cancer is reduced with an increased intake of dietary fiber intake are borne out by the studies reported to date. Lanza (1990) summarized some 48 different studies on the relationship of dietary fiber to colon cancer—38 of the studies (79%) showed a positive inverse relationship, 7 (15%) showed no significant association, and 3 (6%) showed a negative or direct relationship. In effect, some 94% of these studies, which include ecological studies (capable of examining the influence of dietary changes over a significantly long time span) and case control studies, show a positive or neutral effect.

Fiber, Constipation, and Hemorrhoids

One of the "dietary fiber deficiency diseases" that everyone can readily (and sometimes painfully) relate to is constipation. Judging from the relative amount of television, radio, and magazine advertising, as well as the extensive variety and quantity of laxatives available at the retail level, it is obvious that our individual uncomfortable experiences are quite widespread. Increased dietary fiber intake, in particular the insoluble dietary fiber not readily fermented in the large intestine and having high water holding capacity, reduce the incidence of constipation and its side effects.

For insoluble dietary fiber, the physical makeup of the fiber can dramatically affect the action of the fiber on laxation. For example, grinding a coarsely (744 μm) ground fiber down to a fine (173 μm) ground fiber results in a dramatic 40% increase in body transit time (Wrick et al. 1983). Stephen (1985) concluded that cereal brans were more effective than fruits and vegetables for relieving constipation.

Hemorrhoids are believed to occur at least in part from a continually or frequently constipated condition. In particular, constipation causes excess straining, combined with significant shearing forces during the fecal elimination process, which damages the anal area. Burkitt (1985) hypothesized that the excess straining and the increased hardness of the fecal content resulted from low-fiber intake, especially a low intake of cereal fiber in the diet. Neither significant amounts of epidemiological data nor data from designed studies have been accumulated regarding the effect of a high-fiber intake on reducing hemorrhoids. Nonetheless, the reduced strain resulting from softer stools, as discussed earlier, should reduce the potential for anal damage and, if hemorrhoids are present, reduce discomfort during defecation.

Fiber and Crohn's Disease

Burrill Crohn and colleagues first described a disease condition called regional ileitis in 1932 (Rosenberg 1985). After this description, reports of similar conditions in various portions of the large and small intestines expanded the term Crohn's disease to mean regional enteritis. Crohn's disease now refers to any subacute, chronic inflammation, cause unknown, involving any part of the alimentary tract, but particularly the intestinal tract. This typically results in abdominal pain, mild diarrhea, and low grade fever, often accompanied by weight loss.

Although the cause of Crohn's disease is unknown, it appears to have some immunological aspects relative to its occurrence. Consumption levels of dietary fiber, per se, do not appear to have an impact upon the disease. Heaton (1985) maintains, however, that all studies agree that patients with Crohn's disease have been habitually heavy users of refined sugar, typically combined with a low intake of fruits and vegetables. While conjecturing that higher sugar levels might promote the survival of certain intestinal bacteria, he also concedes that diets high in sugar and poor in fruits and vegetables are probably suboptimal in some vitamins and minerals.

Fiber and Diverticular Disease

The incidence of diverticular disease (the appearance of outpouches of the mucosa lining of the large bowel through weak spots in the muscle wall of the bowel) appears to coincide with significant drops in dietary fiber intake (Dreher 1987). It is one of the diseases that provides a clear-cut case of cause/effect relationship with dietary changes. The disease is virtually unheard of in developing countries (Painter 1985), while it is quite prevalent in countries that have a greater availability of processed foods. However, there is not total agreement on the relative incidences of the disease, particularly with regard to historical trends. Eastman (1990) points out that some of the extensive differences in findings on the incidence of the disease may be related to the availability of barium enemas, since the disease is hard to diagnose without the aid of this technique.

Generally, diverticulosis is asymptomatic, with the individual never realizing its presence. In cases where it becomes symptomatic, the unfortunate victim suffers intermittent periods of low grade fever and pain, both of which become steady, with the pain often becoming similar to that of appendicitis. Eastman (1990) and Painter (1985) indicate that, in general, dietary fibers are not necessarily effective in treating the ailment, but coarse wheat bran, on the other hand, is very effective.

Fiber and Ileitis

Ileitis is a common designation given to Crohn's disease when it is confined to the ileum. The earlier section on Crohn's disease discusses the effects of dietary fiber on the disease state.

Fiber and Irritable Bowel Syndrome

Irritable Bowel Syndrome, often referred to as spastic colon, causes abdominal pain (sometimes severe when associated with diarrhea) that results from contractions of the muscle tissue of the colon. The contractions and pain appear and dissipate in a somewhat cyclic fashion. Treatment of the disease with high-fiber diets, usually in the form of wheat bran, has met with mixed success. While it is difficult to prove that higher fiber produces a dramatic curative effect, there are definitely some patients who suffer from the syndrome who have benefited from the changed diet regimen (Harvey 1985).

Fiber and Varicose Veins

Burkitt (1985) reports that patients in communities with low incidences of varicose veins invariably pass large, soft-consistency stools, whereas those patients with small volume, firm-consistency stools have higher rates of incidence of varicose veins. Burkitt acknowledges that the evidence is not statistically validated and is in the form of personal impressions, but he nonetheless believes that it should carry considerable weight because of the abundance and consistency of occurrence and because it reflects the long experience of competent observers. Straining during defecation is postulated to be the most important single cause of this illness; it is brought on by raised intraabdominal pressures resulting from the strain, much as hemorrhoids are brought on by constipation. As discussed in the section on constipation, insoluble dietary fiber, particularly coarsely ground insoluble dietary fiber, is a potential treatment for varicose veins.

Potential Adverse Effects of Increased Dietary Fiber Consumption

With any good thing, there is always the concern that an excess of it will be detrimental. Such is the case with dietary fiber. The most common adverse effect of an excess of dietary fiber is that typically experienced by individuals who are trying to increase their daily intake level too quickly. The results of this rapid increase are often substantial discomfort (both physically and socially) resulting from abdominal pressure and increased flatulence. These symptoms subside, however, as the body adapts to a higher intake level, and as the new level becomes a part of the normal diet. Of longer term concern is the question of whether or not increased dietary fiber intake levels result in vitamin or mineral depletion in the body. Some of the unique properties of dietary fiber that are effective in improving other health states may also lead to a concern, with regard to unacceptably low levels of dietary intake or depletion of these nutrients.

Dietary fiber, particularly insoluble dietary fiber, is a highly water-absorptive bulking agent. As such, it can dilute mineral concentrations in the digestive system while at the same time decrease the total digestive time, due to the decreases in intestinal transit time that occur. By definition, dietary fiber resists digestion and, as a result, may inhibit uptake of vitamins or minerals entrained within its bulk. Furthermore, dietary fiber can act as a weak cation exchanger that associates with and weakly binds minerals in the food (Kay 1982).

Study results regarding vitamin or mineral depletion, with increased dietary fiber intake, are equivocal—the results of one study often conflicting with the results of another. Generally, there appears to be no significant change in vitamin or mineral balance unless the intake levels of dietary fiber are excessively high or the quantity of dietary fiber consumed undergoes a dramatic increase in a relatively brief period of time. Vegetarians, who traditionally consume high levels of dietary fiber on a daily basis, do not have blood mineral levels significantly different than nonvegetarians. Dreher (1987) provides an excellent tabulation and summary of the results of many of these studies conducted over the past 15 years.

That the results of some studies are in conflict with others may possibly be explained by the weak cationic exchange nature of dietary fibers. Many dietary fiber sources are also excellent sources of minerals, particularly micronutrient minerals, such as iron, copper, zinc, chromium, and manganese. If the subjects ingesting the dietary fiber are deficient in some of the micronutrients while having adequate stores of others, it may well be that, with the availability of the mineral in the diet that is in short bodily supply, an exchange takes place. The mineral in short supply might be taken up by the body and the cation exchange site it occupied in the dietary fiber would then be filled with the cation of a mineral that is present in the body in relative excess. The mineral in relative excess would subsequently be excreted with the fecal waste. Future research efforts will be needed that focus on the entire diet and, to the extent possible, seek to understand the balances among all the constituents, both in the diet and in the test subjects.

RECOMMENDED LEVELS OF DIETARY FIBER CONSUMPTION

When any new scientific subject emerges that is sufficiently unprecedented to attract an enduring group of adherents and is sufficiently open ended so as to leave all sorts of problems for the group of adherents to investigate in the future (in other words, a paradigm [Kuhn 1970; Trowell 1985]), there is substantial resistance to making definite recommendations based on the knowledge gained to date. Because there is uncertainty in the results of various studies, and there is uncertainty in the interpretation of those results, it is obvious to those involved that there remains a great deal more to be investigated. Therefore, it is difficult for those involved to reach an individual conclusion, much less a common group consensus, concerning what the appropriate recommendation should be. The paradigm of dietary fiber is no different than other paradigms, driving investigators to explore physiological and dietary relationships in unprecedented detail. Combine this with the variability inherent in research of this type (that is, coronary heart disease can be only indirectly related to dietary fiber consumption on the basis of other risk factors that exhibit considerable variability in response to dietary fiber intake themselves), and it is very difficult for an individual scientist or group of scientists to stake their reputations on a recommendation that may in the future prove to be inaccurate or wrong. As was noted several times previously, there is still much to be done to fully elucidate all the effects of various dietary fiber levels on the human system.

Nevertheless, even in the midst of a rapidly evolving information base, the scientific community owes it to the rest of society, which supports their work and of which they are a part, to recommend changes when the potential benefits to society are great. The potential benefits of increased dietary fiber consumption are substantial, and the relative risks are very low. A lack of recommendations on consumption levels in specific numerical terms is difficult for consumers at large to understand, since a numerical value is something that they can measure against. Furthermore, society at large tends to lose faith in a particular segment of the scientific community if there is substantial unresolved, but highly vocalized differences in recommendations. Therefore, it behooves the individual scientists and professional groups working in the area of dietary fiber to work to define specific dietary recommendations relating to their area of expertise.

Currently, a number of recommendations regarding dietary fiber intake have evolved and surfaced, or are evolving. While there is not necessarily full consensus on these recommendations, they provide an excellent basis for dietary guidelines for the present, as well as a springboard for future recommendations. A number of these recommendations will be discussed here, along with the rationale for arriving at the recommended level, if this information is available.

As early as 1981, Stephen recommended that a dietary fiber intake of 40 g/day was possible for residents of the United Kingdom and therefore ought to be a reasonable target to achieve in the 1980s. Because cereal fiber was the primary fiber source that had decreased in the U.K. diet over the past century, the recommendation included an emphasis on increasing dietary fiber from cereal sources.

Silman and Marr (1985), desiring to promote a better "western diet," recommended increasing the average daily dietary fiber intake from 21.2 to 29.2 grams. Although they considered their increase in the recommended consumption level of dietary fiber to be too modest (low-intake, still in the high-risk area from a health perspective), they believed that this level of intake was a realistic, acceptable, and adoptable modification of the western diet. About half of the dietary fiber would come from cereals.

Anderson et al. (1990) outline two dietary fiber intake levels suggested by the HCF Diabetes Foundation—one for the general public and the other for individuals under medical supervision for diabetes, obesity, or hyperlipidemia. While the recommendations and corresponding diets were developed to treat diabetes, the authors indicate that they have been used successfully in treating and preventing hyperlipidemia. For the general public, their recommendation is 20 to 35 g/day of total dietary fiber intake (10 to 13 g/1000 kilocalories of food consumed), with approximately one third of that as soluble dietary fiber. For individuals under medical supervision, the respective recommended daily intake levels are 35 to 50 grams (20 to 25 g/1000 kilocalories of food consumed), again, with approximately one third of the fiber as soluble fiber.

The British National Advisory Committee on Nutrition Education (NACNE) recommends that individuals make an effort to increase their level of fiber from cereal grains and raise their total daily intake to 30 g/day (NACNE 1986). The basis of their recommendation was an overall improvement in health, particularly in the areas of gastrointestinal and coronary health.

The Expert Advisory Committee on Dietary Fibre recommended in 1985 that dietary fiber intake should be doubled from its current level of 5.8 to 8.0 g/1000 kilocalories, by increasing consumption of a variety of fiber-rich foods (Health and Welfare Canada 1985). The changes were recommended based on three and possibly four physiological effects: 1) regularizing colonic function, 2) normalizing serum lipid levels, 3) attenuating the postprandial glucose response, and, perhaps, 4) suppressing appetite.

The Federation of American Societies for Experimental Biology recommended in 1987 that dietary fiber intake be adjusted to 10 to 13 g/1000 kilocalories, by utilizing a wide variety of foods naturally rich in fiber. Again, the emphasis on the dietary change was to improve the overall health state of individuals, by addressing the variety of illnesses for which increases in dietary fiber intake has been implicated as a preventative or therapeutic factor.

The Nordisk Ministerrad Standing Nordic Committee on Food (1989) recommends a dietary fiber intake ratio of 12 g/1000 kilocalories in food, for the purpose of overall health improvement.

As a result of the studies linking dietary fiber consumption and diabetes, the American Diabetes Association recommends 25 g of fiber per 1000 kilocalories of food consumed (approximately 40 g/day) (American Diabetes Association 1987).

In 1984, on the basis of studies relating dietary fiber consumption and cancer, the National Cancer Institute recommended the consumption of 25 to 35 g/day of dietary fiber. In 1988 (Butrum, Clifford, and Lanza), this recommendation was modified to 20 to 30 g/day, with an emphasis on complex carbohydrates, from fruits, vegetables, and whole cereal grains.

Numerous other agencies, professional organizations, and individuals have issued recommendations regarding dietary fiber intake. The majority of these recommend increasing the level of total and soluble dietary fiber intake, while at the same time reducing consumption of total calories, especially calories derived from saturated fat.

In summary, it appears at this point in time that the opinion of individuals and consensus groups that have issued specific recommendations is to consume a diet having a level of dietary fiber intake of approximately 30 g/day (12 g/1000 kilocalories), for individuals not requiring special medical supervision.

With the variety of high-fiber products becoming available and the rapidly expanding data base of dietary fiber content of these products, the health conscious consumer should be able to select menus high in dietary fiber but with substantial variety. For example, the approximately 30 g/day of dietary fiber could be obtained by ingesting the following.

	TDF	IDF	SDF
		Total Grams	
2 slices of pumpernickel bread	3.5	1.0	2.5
1 portion of Raisin Bran cereal	5.0	4.0	1.0
1 portion red kidney beans	9.0	7.4	1.6
1 baked potato	3.5	2.0	1.5
1 portion Brussels sprouts	2.0	1.7	0.3
1 peach	2.2	1.3	0.9
1 portion of either dried apricots, prunes, or figs	7	4.5	2.5
	32.2	21.9	10.3

This would also provide a balance between soluble and insoluble dietary fiber.

References

Aberg, H., H. Lithell I. Selinus, and H. Hedstrand. 1985. Serum triglycerides are a risk factor for myocardial infarction but not for angina pectoris. *Atherosclerosis* 54:89–97.

Abraham, S. and M. Nordsieck. 1960. Relationship of excess weight in children and adults. *Public Health Reporter* 75:263–273.

Adams, F. 1939. *The Genuine Works of Hippocrates.* Baltimore: Williams and Wilkins.

American Diabetes Association. 1987. Nutritional recommendations for individuals with diabetes mellitus. *Diabetes Care* 10:126–132.

American Dietetic Association. 1989. Position of The American Dietetic Association: Optimal weight as a health promotion strategy. *Journal of the American Dietetic Association* 89:1814–1817.

Anderson, J.W., D.A. Deakins, T.L. Floore, B.M. Smith, and S.R. Whitis. 1990. Dietary Fiber and Coronary Heart Disease. *Critical Reviews*™ in Food Science and Nutrition 29(2):95–147.

Anderson, J.W., N.J. Gustafson, C.A. Bryant, and J. Tietyn-Clark. 1987. Dietary fiber and diabetes: A comprehensive review and practical application. *Journal of the American Dietetic Association* 87(9):1189–1197.

Anderson, J.W., and A.E. Siesel. 1990. Hypocholesterolemic effects of oat products. In *New Developments in Dietary Fiber: Physiological, Physicochemical, and Analytical Aspects*, ed. Ivan Furda and Charles J. Brine, pp. 17–36. New York: Plenum Press.

Anderson, J.W., and K. Ward. 1979. High carbohydrate, high-fiber diets for insulin treated men with diabetes mellitus. *The American Journal of Clinical Nutrition* 32:2312–2321.

AOAC 1990. Method # 985.29 Total dietary fiber in foods. In *Official Methods of Analysis of the Association of Official Analytical Chemists*, ed. Kenneth Helrich, pp. 1105–1106. Arlington, VA: AOAC.

Asp, N.G., I. Furda, J.W. DeVries, T.F. Schweizer, and L. Prosky, 1988. Dietary fiber definition and analysis. *The American Journal of Clinical Nutrition* 47:688–690.

Bell, L.P., K.J. Hectorn, H. Reynolds, and D.B. Hunninghake 1990. Cholesterol-lowering effects of soluble-fiber cereals as part of a prudent diet for patients with mild to moderate hypercholesterolemia. *The American Journal of Clinical Nutrition* 52:1020–1026.

BeMiller, J.N. 1973. Psyllium seed gum. In *Industrial Gums, Polysaccharides and Their Derivatives.* ed. Roy L. Whistler and James N. BeMiller. pp. 345-354. New York: Academic Press.

Betschart, A.A., M.M. Chiu, C.A. Hudson, and J. Tietyen. 1990. Dietary fiber in cereal products—Analytical constituent or physiologically functional component? Paper read at 75th annual meeting of the American Association of Cereal Chemists, Dallas Texas.

Bright-See, E. and G.E. McKeown-Eyssen. 1984. Estimation of per capita crude and dietary fiber supply in 38 countries. *The American Journal of Clinical Nutrition* 39:821–829.

British Nutrition Foundation. 1990. *Complex Carbohydrates in Foods. The Report of the British Nutrition Foundation's Task Force.* pp. 79–82. New York: Van Nostrand Reinhold.

Brownell, K.D. 1984. New developments in the treatment of obese children and adolescents. In *Eating and Its Disorders*, Vol. 62, ed. Albert J. Stunkard and Eliot Stellar, pp. 175–183. New York: Raven Press.

Burkitt, D.P. 1971. Epidemiology of cancer of the colon and rectum. *Cancer: Diagnosis, Treatment, Research* 28:3–13.

Burkitt, D. 1985. Varicose veins, haemorrhoids, deep vein thrombosis and pelvic phleboliths. In Dietary Fibre, Fibre-depleted Foods and Disease, ed. Hugh Trowell, Denis Burkitt, and Kenneth Heaton, pp. 317–330. London: Academic Press.

Burley, V.J., J.E. Blundell, and A.R. Leeds. 1987. The effects of high and low fibre lunches on blood glucose, plasma insulin levels and hunger sensations. *International Journal of Obesity* 11 (Supplement 2):12.

Burley, V.J., and J.E. Blundell. 1990. Action of dietary fiber on the satiety cascade. In *Dietary Fiber: Chemistry, Physiology, and Health Effects*, ed. David Kritchevsky, Charles Bonfield, and James W. Anderson, pp. 227–246. New York and London: Plenum Press.

Burley, V.J., A.R. Leeds, and J.E. Blundell. 1987. The effect of high and low fibre breakfasts on hunger, satiety, and food intake in a subsequent meal. *International Journal of Obesity (Supplement)* 1:87-93.

Butrum, R.R., C.K. Clifford, and E. Lanza. 1988. NCI dietary guidelines: rationale. *American Journal of Clinical Nutrition* 48:888-895.

Castelli, M.P., and K. Anderson. 1986. A population at risk: prevalence of high cholesterol levels in hypertensive patients in the Framingham Study. *American Journal of Medicine* 80:23-32.

Chen, W.J.L., and J.W. Anderson. 1986. Hypocholesterolemic effects of soluble fiber. In *Basic and Clinical Aspects of Dietary Fiber*, ed. G.V. Vahouny and D. Kritchevsky, pp. 275-286. New York and London: Plenum Press.

Cleave, T.L. 1956. The neglect of natural principles in current medical practice. *Journal of the Royal Navy Medical Service* 42:55-83.

Cleave, T.L., and G.D. Campbell. 1966. *Diabetes, Coronary Thrombosis and the Saccharine Disease*. Bristol: John Wright.

Cummings, J. 1985. Cancer of the large bowel. In *Dietary Fibre, Fibre-depleted Foods and Disease*, ed. Hugh Trowell, Denis Burkitt, and Kenneth Heaton, pp. 161-189. London: Academic Press.

Dreher, M.L. 1987. *Handbook of Dietary Fiber*. New York and Basel: Mercel Dekker Inc.

Eastwood, M. 1990. Fiber and gastrointestinal disease. In *Dietary Fiber: Chemistry, Physiology, and Health Effects*, ed. D. Kritchevsky, C. Bonfield, and J.W. Anderson, pp. 261-271. New York: Plenum Press.

Eder, H.A., and L.I. Gidez. 1982. The clinical significance of the plasma high density lipoproteins. *Medical Clinics of North America* 66:431-440.

Englyst, H.N., H. Trowell, D.A.T. Southgate, and J.H. Cummings. 1987. Dietary fiber and resistant starch. *The American Journal of Clinical Nutrition* 46:873-874.

Federation of American Societies for Experimental Biology. 1987. *Physiological Effects and Health Consequences of Dietary Fiber*. Bethesda: Life Sciences Research Office.

Fehily, A.M., M.L. Burr, B.K. Butland, and R.D. Eastham. 1986. A randomized controlled trial to investigate the effect of a high fibre diet on blood pressure and plasma fibrinogen. *Journal of Epidemiology & Community Health* 40:334-337.

Fehily, A.M., J.E. Milbank, J.W.G. Yarnell, T.M. Hayes, A.J. Kubiki, and R.D. Eastham. 1982. Dietary determinants of lipoproteins, total cholesterol, viscosity, fibrinogen, and blood pressure. *American Journal of Clinical Nutrition* 36:890-896.

Food and Agricultural Organization of the United Nations. 1977. Provisional food balance sheets, 1972-1974 average. Rome.

Forbes, G.B. 1975. Prevalence of obesity in childhood. In *Obesity in Perspective*, Vol. 2, ed. G.A. Bray. DHEW Publication No. (NIH) 75-708, Washington, D.C.: U.S. Government Printing Office.

Gordon, T., W.P. Castilli, M.C. Hjortland, W.B. Kannel, and T.R. Dawber. 1977.

High density lipoprotein as a protective factor against coronary heart disease: the Framingham Study. *American Journal of Medicine* 62:707-714.

Harvey, R.F. 1985. Functional gastrointestinal disorders: irritable bowel and other syndromes. In *Dietary Fibre, Fibre-depleted Foods and Disease*, ed. Hugh Trowell, Denis Burkitt, and Kenneth Heaton, pp. 217-228. London: Academic Press.

Health and Welfare Canada. 1985. *Report of the Expert Advisory Committee on Dietary Fibre to the Health Protection Branch, Health and Welfare Canada.* Ottawa: Minister of National Health and Welfare.

Heaton, K. 1985. Crohn's disease and ulcerative colitis. In *Dietary Fibre, Fibre-depleted Foods and Disease*, ed. Hugh Trowell, Denis Burkitt, and Kenneth Heaton, pp. 205-216. London: Academic Press.

Hipsley, E. H. 1953. Dietary "fibre" and pregnancy toxaemia. *British Medical Journal* 2:420-422.

Holt, S., R.C. Heading, D.C. Carter, I.F. Prescott, and P. Tothill. 1979. Effect of gel fibre on gastric emptying and absorption of glucose and paracetamol. *The Lancet* 2:636-639.

Hubert, H.B., M. Feinleib, P. McNamara, and W.P. Castelli. 1983. Obesity as an independent risk factor for cardiovascular disease: a 26 year followup of participants in the Framingham Heart Study. *Circulation* 67:968-977.

Jenkins, D.J.A., A.L. Jenkins, T.M.S. Wolever, and V. Vuksan. 1990. Fiber and physiological and potentially therapeutic effects of slowing carbohydrate absorption. In *New Developments in Dietary Fiber: Physiological, Physicochemical, and Analytical Aspects*, ed. Ivan Furda and Charles J. Brine, pp. 129-134. New York: Plenum Press.

Jenkins, D.J.A., D.M. Thomas, M.S. Wolever, R.H. Taylor, H. Barker, H. Fielden, J.M. Baldwin, A.C. Bowling, H.C. Newman, A.L. Henkins, and D.V. Goff. 1981. Glycemic index of foods: a physiological basis for carbohydrate exchange. *American Journal of Clinical Nutrition* 34:362-366.

Kannel, W.B. 1986. Hypertension: relationship with other risk factors. *Drugs* 31 (supplement 1): 1-11.

Kannel, W.B., T.R. Dawber, and D.L. McGee. 1980. Perspectives on systolic hypertension: the Framingham Study. *Circulation* 61:1179-1182.

Kannel, W.B., and T. Gordon. 1979 (November). Physiological and medical concomitants of obesity: the Framingham Study. In *Obesity in America*, ed. G.A. Bray. pp. 125-163. Beltsville: United States Department of Health, Education, and Welfare, Public Health Service, National Institutes of Health, NIH Publication No. 79-359.

Kannel, W.B., and P.A. Sytkowski. 1987. Atherosclerosis risk factors. *Pharmacology and Therapeutics.* 32:207-235.

Kannel, W.B., P.A. Wolf, W.P. Castelli, and R.B. D'Agostino. 1987. Fibrinogen and risk of cardiovascular disease. *Journal of the American Medical Association* 258:1183-1186.

Kay, R.M. 1982. Dietary fiber. *Journal of Lipid Research* 23:221-242.

Kiehm, T.G., J.W. Anderson, and K. Ward. 1976. Beneficial effects of a high carbohydrate, high-fiber diet on hyperglycemic men. *American Journal of Clinical Nutrition* 29:895–899.

Kinosian, B.P., and J.M. Eisenberg. 1988. Cutting into cholesterol: cost effective alternatives for treating hypercholesterolemia. *Journal of the American Medical Association* 259:2249–2254.

Koepp, P. and S. Hegewisch. 1981. Effects of guar on plasma viscosity and related parameters in diabetic children, *European Journal of Pediatrics.* 137:31–33.

Kozmin, P.A. 1921. *Flour Milling*. New York: Van Nostrand.

Kritchevsky, D. 1988. Dietary fiber. *Annual Review of Nutrition* 8:301–328.

Kuhn, T.S. 1970. *The Structure of Scientific Revolutions.* Chicago: University of Chicago Press.

Lanza, E. 1990. National Cancer Institute satellite symposium on fiber and colon cancer. In *Dietary Fiber: Chemistry, Physiology, and Health Effects,* ed. D. Kritchevsky, C. Bonfield, and J.W. Anderson, pp. 383–387. New York: Plenum Press.

Levine, A.S., J.R. Tallman, M.K. Grace, S.A. Parker, C.J. Billington, and M.D. Levitt. 1989. Effect of breakfast cereals on short-term food intake. *The American Journal of Clinical Nutrition* 50:1303–1307.

Luft, F.C., J.Z. Miller, R.M. Lyle, C.L. Melby, N.S. Fineberg, D.A. McCarron, M.H. Weinberger, and C.D. Morris. 1989. The effect of dietary interventions to reduce blood pressure in normal humans. *Journal of the American College of Nutrition* 8(6):495–503.

Lyle, R.M., C.L. Melby, G.C. Hyner, J.W. Edmondson, J.Z. Miller, and M.H. Weinberger. 1987. Blood pressure and metabolic effects of calcium supplementation in normotensive white and black men. *The Journal of the American Medical Association* 257:1772–1776.

McCarron, D.A., and C.D. Morris. 1987. The calcium deficiency hypothesis of hypertension. *Annals of Internal Medicine*. 107(6):919–922.

NACNE. 1986. Nutritional guidelines for health education in Britain. *Nutrition Today* 21:21–22.

National Cancer Institute. 1984. *Diet, Nutrition, and Cancer Prevention: A Guide to Food Choices*. Publ. No. (NCI) 85–2711, National Institutes of Health.

National Institutes of Health Consensus Development Panel on the Health Implications of Obesity. 1985. Health Implications of Obesity: National Institutes of Health Consensus Development Conference statement. *Annals of Internal Medicine* 103:147–151.

Nordisk Ministerrad Standing Nordic Committee on Food. 1989. *Nordic Nutrition Recommendation (2nd ed) Report 2.*

Painter, N.S. 1985. Diverticular disease of the colon. In *Dietary Fibre, Fibre-Depleted Foods and Disease*, ed. H. Trowell, D. Burkitt, and K. Heaton, pp. 145–160. London: Academic Press.

Paul, A.A., and D.A.T. Southgate. 1978. *McCance and Widdowson's composition of foods*. London: Her Majesty's Stationery Office.

Prosky, L., and B.F. Harland. 1979. Need, Definition, and Rationale Regarding

Dietary Fiber. Paper read at 93rd Annual Meeting of the Association of Official Analytical Chemists, Washington, D.C.

Rivellese, A., G. Riccardi, A. Giacco, D. Pancioni, S. Genovese, P.L. Mattioli, and M. Mancini. 1980. Effect on dietary fibre on glucose control and serum lipoproteins in diabetic patients. *Lancet* 2:447–450.

Rosenberg, I.H. 1985. Crohn's disease. In *Cecil Textbook of Medicine,* ed. James B. Wyngaarden and Lloyd H. Smith, Jr. pp. 740–748. Philadelphia: W.B. Saunders Co.

Schneeman, B.O., and M. Lefevre. 1986. Effects of fiber on plasma lipoprotein composition. In *Dietary Fiber: Basic and Clinical Aspects,* ed. G.V. Vahouny and D. Kritchevsky, pp. 309–321. New York: Plenum Press.

Schweizer, T.F. 1989. Dietary fiber analysis. *Lebensmittel-Wissenschaft & Technologie* 22:54–59.

Shils, M.E. 1988. Magnesium in health and disease. In *Annual Review of Nutrition,* ed. Robert E. Olson, Ernest Beutler, and Harry P. Broquist, pp. 429–460. Palo Alto, CA: Annual Reviews Inc.

Silman, A., and J. Marr. 1985. A better western diet: What can be achieved? In *Dietary Fibre, Fibre-depleted Foods and Disease,* ed. Hugh Trowell, Denis Burkitt, and Kenneth Heaton, pp. 403–418. London: Academic Press.

Simopoulos, A.P., and T.B. Van Itallie. 1984. Body weight, health and longevity. *Annals of Internal Medicine* 100:285–295.

Simpson, H.C.R., J.I. Mann, R. Chakrabarti, J.D. Imeson, Y. Stirling, M. Tozer, L. Woolf, and T.W. Meade. 1982. Effect of high-fiber diets on haemostatic variables in diabetes. *British Medical Journal* 284:1608.

Simpson, H.C.R., R.W. Simpson, S. Lousley, R.D. Carter, M. Geekie, T.D.R. Hockaday, and J.I. Mann. 1981. A high carbohydrate leguminous fibre diet improves all aspects of diabetic control. *The Lancet* 1:1–5.

Spiller, G.A. 1990. Complexity in the interpretation of data derived from studies of dietary fiber. In *New Developments in Dietary Fiber: Physiological, Physicochemical, and Analytical Aspects,* ed. Ivan Furda and Charles J. Brine, pp. 179–181. New York and London: Plenum Press.

Stephen, A. 1981. Should we eat more fibre? *Journal of Human Nutrition* 35:403–414.

Stephen, A. 1985. Constipation. In *Dietary Fibre, Fibre-depleted Foods and Disease,* ed. Hugh Trowell, Denis Burkitt, and Kenneth Heaton, pp. 133–144. London: Academic Press.

Storck, J. and W.D. Teague. 1952. *Flour for Man's Bread, A History of Milling.* Minneapolis, MN: University of Minnesota Press.

Story, J.A. 1980. Dietary fiber and lipid metabolism. *Proceedings of the Society for Experimental Biology and Medicine* 180:447–452.

Trowell, H.C. 1960. *Non-Infective Disease in Africa.* London: Edward Arnold.

Trowell, H. 1972a. *Revue Europeene d'Etudes Cliniques et Biologiques* 17:345–349.

Trowell, H. 1972b. Ischemic heart disease and dietary fiber. *The American Journal of Clinical Nutrition* 25:926–932.

Trowell, H. 1972c. Crude fibre, dietary fibre and atherosclerosis. *Atherosclerosis* 16:138–140.

Trowell, H.C. 1973. Dietary fibre, ischaemic heart disease and diabetes mellitus. *Proceedings of the Nutrition Society* 32:151–157.

Trowell, H.C. 1974. Diabetes mellitus death rates in England and Wales, 1920–70, and food supplies. *The Lancet* 2:998–1001.

Trowell, H.C. 1990. Fiber depleted starch and NIDDM. In *Dietary Fiber: Chemistry, Physiology, and Health Effects*, ed. D. Kritchevsky, C. Bonfield, and J.W. Anderson, pp. 283–286. New York: Plenum Press.

Trowell, H., D. Burkitt, and K. Heaton. 1985. Definitions of dietary fibre and fibre-depleted foods. In *Dietary Fibre, Fibre-Depleted Foods and Disease*, ed. Hugh Trowell, Denis Burkitt, and Kenneth Heaton, pp. 21–30. London: Academic Press.

Trowell, H., E. Godding, G. Spiller, and G. Briggs. 1978. Fiber bibliographies and terminology. *The American Journal of Clinical Nutrition* 31:1489–1490.

United States Department of Agriculture, Agriculture Research Service. 1990a. Young women who avoid eating whole grains, vegetables, nuts, and seeds may be losing more than iron during menstruation. In *Quarterly Report of Selected Research Projects*, July 1 to September 30:4.

United States Department of Agriculture, Agriculture Research Service. 1990b. Middle age diabetes doesn't happen overnight. In *Quarterly Report of Selected Research Projects*, July 1 to September 30:4.

United States Department of Health and Human Services. 1988. *The Surgeon Generals Report on Nutrition and Health*, DHHS (PHS) Publication number 88-50210. Washington, D.C.: Government Printing Office.

Van Soest, P.J., B.A. Lewis, and J.B. Robertson. 1988. Fiber in the diet of some rural populations in the Peoples Republic of China. *The FASEB Journal* 2:A1081.

Vega, G.L., E. Grosezek, R. Wolf, and S.M. Grundy. 1982. Influence of polyunsaturated fats on composition of plasma lipoproteins and apolipoproteins. *Journal of Lipid Research* 23:811–822.

Walker, A. and D. Burkitt. 1985. Appendicitis. In *Dietary Fibre, Fibre-Depleted Foods and Disease*, ed. H. Trowell, D. Burkitt, and K. Heaton. pp. 191–203. London: Academic Press.

Wilhelmsen, L., K. Svardsudd, K. Korsan-Bengtsen, B. Larsson, L. Welin, and G. Tibblin. 1984. Fibrinogen as a risk factor for stroke and myocardial infarction. *New England Journal of Medicine* 311(8):501–506.

Wolever, T.M.S. 1990. Dietary fiber in the management of diabetes. In *Dietary Fiber: Chemistry, Physiology, and Health Effects*, ed. D. Kritchevsky, C. Bonfield, and J.W. Anderson, pp. 283–286. New York: Plenum Press.

Wrick, K.L., J.B. Robertson, P.J. Van Soest, B.A. Lewis, J.M. Rivers, D.A. Roe, and L.R. Hackler. 1983. The influence of dietary fiber source on human intestinal transit and stool output. *Journal of Nutrition* 113:1464–1479.

2

Development of the Association of Official Analytical Chemists Method for Total, Soluble, and Insoluble Dietary Fiber

INTRODUCTION

A.R.P. Walker (1947) was one of the first contemporary researchers in the fiber field to report on the profound implications of ingesting dietary fiber. But it remained for T. L. Cleave (1966), father of the dietary fiber hypothesis, to observe and report that many of the most common diseases associated with western civilization were rare in the less economically developed countries. Inhabitants of these countries experienced a comparably low prevalence of certain diseases in both black and white subpopulations who ingested similar diets, while white subpopulations who ingested a more refined diet experienced a significantly higher incidence. Cleave intuitively recognized that the important factor common to all situations in which these diseases were most frequently observed was the relatively high consumption of refined carbohydrate foods, when compared to the consumption patterns of subpopulations that were not exhibiting these diseases. His great vision was the realization that the human body must be poorly adjusted to modified foods—namely, refined carbohydrates, sugar, and white wheat flour.

Cleave spent his life gathering evidence that supported the view that a wide variety of diseases were caused by consuming foods that contained refined carbohydrates. These diseases included appendicitis, constipation, coronary heart disease, dental caries, diverticulitis, diabetes, obesity, peptic ulcer, maturity onset diabetes, and varicose veins. These ideas were perpetuated by advocates such as Burkitt, Trowell, and Painter, who first suggested the importance of a diet rich in plant foods and who demonstrated that the common thread in the diet was a deficiency of fiber, rather than the presence or excess of refined carbohydrates. Independent of the work of these medical practitioners, Van Soest in the United States and Southgate in the United Kingdom were devising techniques to analyze fiber and its components in foods.

The methodology for determining dietary fiber has been developed over a long period of time and has been hampered by a lack of a definition of fiber. Part of the complication has arisen because fiber is not a single entity, but is in fact a mixture of many complex organic substances. Although many perceived the need for a definition and method for dietary fiber for many years, this need was brought to the attention of a broad segment of the scientific community by the publication of a paper by Burkitt et al. (1972), who represented dietary fiber as a food ingredient that would cure or reduce the incidence of a number of chronic degenerative diseases. By the time a U.S. government publication (U.S. Department of Agriculture and Department of Health and Human Services 1980) recommended Americans should "eat foods with adequate starch and fiber," the U.S. Food and Drug Administration (FDA), through the Association of Official Analytical Chemists (AOAC), had begun working on an acceptable definition and method for dietary fiber. It is only in the past 15 years that dietary fiber has been fully recognized as an important component of food—at the "Marabou Food and Fiber Symposium" (1976), held in Sweden; the "Symposium on Role of Dietary Fiber in Health," held in Bethesda, Maryland, March 1977 (Roth and Mehlman 1978); the European Economic Community Working Group (Theander and James, 1979); the George Washington University Fiber Symposium (Vahouny and Kritchevsky 1982, 1984, and 1988); the International Symposium on Dietary Fibre, held at the University of Reading in March (1982a); the International Symposium on Fiber in Human and Animal Nutrition at Massey University, New Zealand, in May 1982; Symposium on Dietary Fiber with Clinical Aspects, Helsingor, Denmark, 1986; Symposium on Dietary Fiber, American Chemical Society, 1989; and Fibre 90, Chemical and Biological Aspects of Dietary Fibre, Royal Society of London, 1990.

Measuring dietary fiber was to be the first step in identifying the quantity of fiber in the diet. Subsequently, the physical and physiological effects of dietary fiber, including its interactions with other nutrients, would be assessed. This chapter tells how the definition of dietary fiber and the method for its determination were developed.

Dietary fiber's first definition by Hipsley (1953), which included cellulose, hemicellulose, and lignin, has been broadened by some, to include soluble substances, such as gums, modified celluloses, mucilages oligosaccharides, and pectins (Trowell et al. 1976). This broader definition has been suggested because additional nondigestible carbohydrate components have been discovered to have physiological actions. Components such as gums, mucilages, pectins, and lignin may speed gastric emptying time and bind bile acids. Other components, such as alginic acid and cellulose, increase the liver cholesterol content in cholesterol-fed rats, while lignin, which increases steroid excretion, may bind trace elements. Pectin, on the other hand, has been shown to have a cholesterol-lowering effect. Measuring *crude fiber*, "the residue of plant food left after extraction with solvent, dilute acid, and dilute alkali" (Van Soest and McQueen 1973), does not include these additional fiber components and, therefore, is of little value to the clinician.

ASSOCIATION OF OFFICIAL ANALYTICAL CHEMISTS (AOAC) INVOLVEMENT

In 1975, the AOAC published as official first action, a method proposed by Van Soest (Van Soest 1963) for acid detergent fiber (ADF) that measured cellulose and lignin, while a neutral detergent fiber (NDF) method was adopted as an official method by the American Association of Cereal Chemists (AACC), as an alternative to the crude fiber method (Van Soest and Wine 1967). The NDF method is a rapid chemical method that gives higher estimates of fiber than the crude fiber method because of more complete recoveries of cellulose, hemicellulose, and lignin. Since complete removal of starch by conventional methods, such as NDF, is difficult in some food samples, the method was modified in 1978, by adding an alpha-amylase treatment to remove residual starch (AACC 1978). Neither the ADF nor the NDF methods include the more recently discovered physiologically active components that have been encompassed by the term "dietary fiber," nor have these methods been subjected to extensive collaborative studies on diverse foods and feeds. Approximately ten years ago, at the dietary fiber workshop of the XI International Congress of Nutrition, in Rio de Janeiro, Brazil (Spiller and Kay 1979), several scientists, active in dietary fiber research, met to discuss the current state-of-the-art, with respect to the definition and methodology of dietary fiber. The following recommendations resulted from that meeting:

1. Both the water-insoluble and the water-soluble carbohydrate polymers that are not hydrolyzed by human digestion enzymes should be included in the overall definition of fiber.

2. Crude fiber bears no consistent quantitative relationship to dietary fiber in a food, and the use of crude fiber should be abandoned.
3. The NDF method is (currently) acceptable for routinely determining insoluble fiber. For foods containing starch, hydrolysis of the amylose fraction must precede the neutral detergent steps.
4. Determination of water-soluble components (gums, mucilages, pectins) must be included in routine methods for dietary fiber.
5. Enzymatic methods may have widespread applicability in fractionation procedures.
6. Although rapid quantitative methods are acceptable for routine use, such as industry quality control, more tedious methods are essential for defining the chemical characteristics of specific fibers with known physiological effects.
7. Food and feed standards should be made available for quality control and for further characterization of their physicochemical properties. Standards should include defined fiber isolates (cellulose, lignins, pectins), as well as chemically characterized native food sources.

All these recommendations have been acted upon to some extent in subsequent years.

At the 93rd Annual Meeting of the Association of Official Analytical Chemists in Washington, D.C., Prosky and Harland (1979) announced their intention to seek a definition and a method for analyzing total dietary fiber (TDF), which would then be subjected to collaborative study. In the spring of 1981, at an AOAC Workshop in Ottawa, Canada (Prosky 1981), Asp, Baker, Heckman, Southgate, and Van Soest reported on their fiber research. From these reports, the scientists present concluded that two methods for determining total dietary fiber should be developed: (1) a rapid gravimetric method (a modification of the enzymatic methods developed by Asp et al. 1983; Schweizer and Würsch 1979; and Furda 1981); and (2) a more comprehensive method, such as a modification of Southgate's procedure (Southgate 1969), to determine the individual dietary fiber components. In the initial method, the sum of the soluble and insoluble polysaccharides and lignin would be defined and measured as a unit as Total Dietary Fiber (TDF). In the second method, each of the specific components of TDF would be identified and measured separately (Prosky and Harland 1981). It was recognized that the needs of food scientists, who were developing food products, were different from those of physiologists, who were primarily interested in identifying the fiber fractions that most consistently elicited physiologic responses related to human health. Further, it was suggested that the latter method might serve to verify the former method.

By 1981, at the 95th Annual AOAC Meeting in Washington, D.C., following a written inquiry by Prosky and Harland (FDA), more than 100 persons had expressed an interest in the area of dietary fiber, by suggesting definitions of dietary fiber and a preferred method for analysis. Most respondents preferred the definition of Trowell (1974): "Dietary fiber consists of remnants of the plant cells resistant to hydrolysis by the alimentary enzymes of humans." This definition was later modified by Trowell, Southgate, and others (Trowell et al. (1976)) to include cellulose, gums, hemicellulose, lignin, oligosaccharides, pectins, and waxes. The preferred method for analyzing dietary fiber at that time was a modification of Southgate's procedure (Southgate 1969). Because of the complexities and time involved with this procedure (even though the method gave the most complete answers, with regard to the makeup of dietary fiber), there was a desire for an intermediate method, one that would include analyzing the insoluble and soluble fractions that had been lost in previous methods (acid detergent fiber (ADF), neutral detergent fiber (NDF), and yet simpler than the Southgate method. Fiber analysts sought one that could be carried out in most general chemical laboratories. Thus, the collaborative study for the enzymatic-gravimetric method, developed by Prosky, Asp, Furda, De Vries, Schweizer, and Harland, was submitted for collaboration in February 1982. At the AOAC Annual Meeting in October 1982, Prosky and Harland (1982) reported on the progress of the collaborative study for total dietary fiber. They published two papers, reporting the results of the collaborative work (Prosky et al. 1984; 1985). Their enzymatic-gravimetric method was accepted by the AOAC as official final action. Two other papers published subsequently (Prosky et al. 1988; 1990) indicate that the original method, with some minor modification, may be applicable for determining soluble and insoluble dietary fiber. In the near future, a number of modifications in the AOAC methods will be tested collaboratively in a series of studies. These include Li and Andrew's modification, which omits the protease step (Li and Andrews 1988), as well as Lee's modification, which changes the buffer system from phosphates to morphilino ethanesulfonic acid and tris hydroxymethyl aminomethane (MES-TRIS) (Personal Communication). There are also plans to collaborate Theander's methods (Theander and Westerlund 1986) and the Englyst procedure (Englyst and Cummings 1988). Another method in the planning stages for collaborative study is the urea enzymatic dialysis (UED) method of Jeraci et al. (1989). The method of Mongeau and Brassard (1986) is also being considered as an alternative procedure to the current official AOAC-accepted method for total dietary fiber. In this method, one measures the neutral detergent fiber (water-insoluble) and the water-soluble fiber and adds them, which gives the total dietary fiber. It is also an enzymatic-gravimetric method of analysis.

HISTORY OF THE DEVELOPMENT OF DIETARY FIBER METHODOLOGY

In 1806, Einhof obtained fiber values on a number of feed samples by simply macerating the test portion and subsequently extracting with hot water. Crude fiber determinations obtained by the sequential extractions of plant foods by ether, acid, and alkali also had been developed in the early nineteenth century. In 1887, this method for crude fiber was adopted by the Association of Official Agricultural Chemists (predecessor to the AOAC) (Browne 1940). The analysis for crude fiber was favored at that time because it involved the easily obtained reagents H_2SO_4 and NaOH and was reasonably reproducible. However, major fractions of the plant fiber, such as hemicellulose (ca. 85% lost), lignin (ca. 90% lost), and even some of the cellulose, were lost due to their solubility in acid or in alkali. This crude fiber method was the analytical method of choice for the analysis of foods and feeds for approximately 40 years. From these analyses, a large data base of crude fiber values was developed.

Much later, when it became clear that ingesting particular components of the fiber fraction of foods or that foods containing particular fiber components indeed aided in managing certain diseases (that is, diverticular disease and diabetes (Painter 1975; Anderson 1982)), measuring those components became important to clinicians. First, clinicians wanted to measure fiber more accurately, to quantitate how much was in the diet. Second, they wanted to see how the human diet had changed over the years, with regard to fiber intake, and, finally, they began to consider essential a definition of fiber's action in metabolic processes. The term dietary fiber came to be generally applied to the fraction that could elicit physiological response, as opposed to the older term, crude fiber, of the classical Weende proximate system of food analysis.

Editions of McCance and Widdowson's Composition of Foods from 1940 to 1960 began first to report the direct analysis of available carbohydrates, sugars, dextrins, and starch, and second, but more recently, the direct analysis of unavailable carbohydrates. These were defined as the sum of polysaccharides and lignin, which are not digested by the endogenous secretions of the gastrointestinal tract. The polysaccharides were of a number of types, including pectic substances, hemicellulose, and cellulose, and the non-carbohydrate was lignin. Taken together, these fractions were then called dietary fiber (Paul and Southgate 1978). Trowell (1972a) had suggested the term dietary fiber, preserving the word fiber from crude fiber, but changing the adjective from crude to dietary, to show the inadequacy of the crude fiber determination, as well as the necessity of analyzing all of the unavailable carbohydrates (cellulose, gums, hemicellulose, pectins) and lignin in the diet (Trowell 1972bc). In a letter dated December 4, 1980, Trowell stated, "For food labelling and food tables it is appropriate to report 'total dietary fibre,' perhaps better called 'dietary fibre.' This is the sum of the undigested-by-enzymes primary structural fractions, hemicellulose, cellulose, lignin, and pectin. If other fractions are proved to be of importance in health or disease, they can be added subsequently." A copy of this letter was sent to Sir Douglas Black, President of the Royal College of Physicians and Chairman of their Working Party. Trowell further stated in a letter dated April 8, 1981, that, "Objections can be raised to any terminology and definition, especially those put forward at the initial stage of any enquiry. Medical history records that the initial terminology in any field after a certain stage is seldom changed, it is the definition that usually alters with progressive clarification. The original term usually sticks. The vitamin group are not vital amines—the term remains, but the definition changes. Objections can be raised against almost every terminology and definition. Quite frankly I am not certain that John Cummings' attempt to define dietary fibre as the nonstarch polysaccharides of the diet (Proc. Nutr. Soc. 1981, 40:7) is an improvement, for one has to state if glycogen is excluded and why lignin is included. I have always felt there must be a chemical definition of the principal constituents, then analyses; subsequently we must study the physico-chemical relationships. These cannot be studied first, so I feel that we should not emphasize them at the earliest stage, or make them the basis of any definition, yet Heaton (Am. J. Clin. Nutr. 1979, 32:2373) almost suggests this."

The essence of all of the compounds included in the term "dietary fiber" was that they were not digested by the alimentary enzymes of man. In the beginning, it was assumed that all of these compounds had the plant cell wall as their origin. Later, the term "dietary fiber" included *all* undigested polysaccharides, such as storage polysaccharides present in leguminous seeds (Trowell 1974). The term also did not include polysaccharides present in some food additives, such as plant gums, algal polysaccharides, pectins, modified celluloses, and modified starches. The definition of dietary fiber was extended to include all polysaccharides and lignin in the diet that was undigested by endogenous secretions of the digestive tract (Trowell et al. 1976; Trowell 1976), simply because cell-wall polysaccharides usually cannot be differentiated from other nondigestible polysaccharides originally present in plant, nor those incorporated as food additives.

It was Heaton (1983) who also urged the use of three different descriptions of fiber—botanical, chemical, and physiological—to obviate using a restricted definition (Southgate 1982) in determining the physiological effects of dietary fiber.

The revised definition is generally accepted. It should be noted, however, that a very small amount of the dietary fiber component of most diets comes from food additives. In 1974, the Kellogg Company held its first fiber symposium; its second was in 1977. Chemists, food scientists, physicians, and epidemiologists discussed their experimental observations. A symposium held at the Royal College of Physicians of London was concerned with the relationship of dietary fiber ingestion and metabolic disease. There was some discussion of methodology and the measurement of dietary fiber (Baird and Ornstein 1981). In 1977, the European Economic Community and the International Agency for Research on Cancer, a part of the United Nations World Health Organization, held a workshop in Lyon, France. Some of the recommendations pertaining to dietary fiber analysis made by the representatives of the European countries and the United States were: (1) crude fiber should be replaced by dietary fiber analyses, (2) dietary fiber analyses in food should be based on a thorough understanding of the molecular structure of the complex carbohydrates, (3) all properties of dietary fiber, such as solubility, density, hydration, ion-exchange capacities, particle size, and susceptibility to microbial fermentation, should be researched, and (4) fat extraction should be a prerequisite to dietary fiber analyses. Guidelines for analyses of nine different food samples were set, with plans to assemble again in Cambridge, England, in December 1978, to discuss progress to date.

These early conferences represented a commitment to the tremendous effort that would be required to change concepts, approaches, and, above all, the food and feed fiber data bases. There had been acceptance of the fiber values determined by the crude fiber method, but it was time to discard them in favor of some new values (that is, TDF, fractions of which had been shown to have a role in managing diabetes and diverticular disease).

One approach was to name the fiber fraction based on its method of preparation (that is, ADF or NDF). If one were to reflux foods or feeds for 1 h in a water solution of 1 N sulfuric acid and 2% cetyltrimethylammonium bromide (Van Soest 1973), the residue consisting mainly of cellulose and lignin, plus the Mailliard products of heating and cooking, was then called acid detergent fiber. In this procedure, hemicelluloses were lost, as well as the water-soluble portions of the fiber. An advantage of this procedure was that further analysis for cellulose and lignin could be performed on the residue.

The chemical analysis that is used to measure neutral detergent fiber consists of boiling plant materials with neutral sodium lauryl sulfate and ethylene diamine tetracetic acid (EDTA). Lignin, cellulose, hemicellulose, and such cell-wall components as cutin, minerals, and protein are salvaged, although soluble carbohydrates, lipids, and pectins are lost.

Reviews of dietary fiber analytical methods have appeared in scientific and medical literature (Selvendran and Dupont 1984; Asp and Johansson 1984). Recently, a paper was published that gave more than a comprehensive review; it gave an assessment of some of the more promising methods (Selvendran et al. 1989). Methods for determining dietary fiber may be divided into three categories:

1. Gravimetric methods;
2. Colorimetric methods; and
3. Gas-liquid chromatographic methods (GLC) or other methods that measure monomeric composition (HPLC).

Gravimetric methods measure, by weight difference, the insoluble residue remaining after chemical or enzymatic solubilization of non-fiber components. The non-fiber components are usually starch and protein. Recently, this methodology has been extended to the measurement of soluble and insoluble dietary fiber (Prosky et al. 1988; Prosky 1991). One of the more popular gravimetric methods used is to estimate the weight of the residue, following detergent extraction of the product. Van Soest (1963), in his development of the ADF method, tried to reduce the blank caused by the protein. This problem was solved using cetyltrimethylammonium bromide (CTAB) in sulfuric acid, but the treatment resulted in the loss of residues of pectin (Belo and de Lumen 1981) and hemicellulose (Morrison 1980).

Van Soest and Wine Method

In the neutral detergent fiber method (NDF), Van Soest and Wine (1967) boiled the test portion in a neutral, buffered solution of sodium dodecyl sulfate (SDS) and EDTA. The hemicellulose was left in the fiber residue, while pectins were extracted with EDTA. The neutral detergent system solubilized protein efficiently, and some of the fat but starchy materials caused filtration problems with concomitant and erroneously high-fiber results. Several modifications have been introduced in order to improve starch removal. The American Association of Cereal Chemists (AACC) adopted a method for insoluble dietary fiber in cereal products in which NDF residue is treated with amylase, followed by starch removal by washing (AACC Method 32-20). Robertson and Van Soest (1977) found that by including amylase in the NDF reagent they could solve the filtration problem. It was further reported by McQueen and Nicholson (1979) and Schweizer and Würsch (1979, 1981) that preincubation with amylase would ease the filtration step. The detergent methods discussed here were originally designed for analyzing fiber in feeds and forages, but they have been applied to human foods as well. Their main advantages are simplicity and speed of determination. Their primary disadvantage is that soluble components are neither determined nor easily isolated from the filtrate. The NDF methods can be useful in the routine analysis of fiber in many foods, if a relationship between NDF and total dietary fiber in the food analyzed can be established.

Prosky, Asp, Furda, DeVries, Schweizer, and Harland Method

The AOAC-TDF method (Prosky et al. 1984, 1985) was developed and based on the experience of three groups of workers (Asp et al. 1983; Furda 1981; Schweizer and Würsch 1979) and has been extended through collaborative trials to include the measurement of soluble dietary fiber (SDF) and insoluble dietary fiber (IDF) (Prosky et al. 1988). The total dietary fiber (TDF) method has been accepted as official by the AOAC and has been accepted by the U.S. Food and Drug Administration as the method of choice for labeling and for any other purposes for which TDF is declared. The insoluble and soluble dietary fiber methods, which are modifications of the TDF method, are in the process of being proposed as official methods of the AOAC. They have been used for some time by many investigators throughout the world for analyses.

The TDF method has been approved as the legal or recommended procedure for the analyses of foods in at least ten countries, including the United States, the Nordic countries, West Germany, and Switzerland, and has recently undergone additional validation studies (Schweizer et al. 1988; Rabe 1987; Rabe et al. 1988). Additionally, the AOAC method is now listed in the Australian Food Standards Code and is often included on food labels. It is at the present time the most widely used fiber method because it meets the requirements for food labeling and can be used for associated quality control, yet is fairly rapid. It is less rapid than the crude fiber and NDF methods, but the results are more meaningful. More than one hundred papers have reported data, using this relatively new AOAC-TDF method for the fiber content of foods. This will help to construct TDF composition tables, for foods using data derived from many different laboratories throughout the world. An objection raised against the AOAC method and other enzymatic-gravimetric procedures is that the chemically undefined fraction measured as dietary fiber has no predictive value for physiological effects (Englyst et al. 1988; Englyst and Cummings 1988). These criticisms appear to be unfounded when one considers what can reasonably be expected from a routine measurement.

None of the currently accepted methods for proximate analysis of food—Kjeldahl nitrogen × 6.25 for protein, the gravimetric determination of fat and total ash, and the direct analysis of sugar and starch—predict physiological effects. There are many nutritional and dietary factors that influence the physiological activity of the macronutrients. The gravimetric residues from the TDF, IDF, and SDF procedures can be further investigated for their monosaccharide, uronic acid, or Klason lignin components in order to obtain the same detailed information derived from some of the chemical procedures described in the following pages. This has been repeatedly demonstrated by different investigators who utilized different gravimetric procedures, including the AOAC method (Schweizer and Würsch 1979; Marlett and Navis 1988; Nyman et al. 1987).

Another objection raised regarding the AOAC method is that resistant starch (RS) is included in the dietary fiber fraction. There is some disagreement about whether or not resistant starch should be included in the dietary fiber fraction (Englyst et al. 1987; Asp et al. 1988). This is a form of starch that has been shown to be neither hydrolyzed nor absorbed in the small intestine (Englyst and Cummings 1985; Bjork et al. 1986). Resistant starch is sometimes formed during food processing, usually at low levels. This starch is not digestible, but readily fermented by the intestinal flora, as are many other fiber polysaccharides.

A recent study has shown that all starch included in the AOAC dietary fiber residue of mixed diets escapes digestion and absorption in the small intestine (Schweizer et al. 1990), justifying its inclusion in the dietary fiber fraction. Englyst et al. (1987) argue that variations in processing may cause variations in the amount of RS formed, thus making it difficult to prepare food composition tables. This is not a problem unique to dietary fiber. Existing food tables list the vitamin content for raw and variously processed vegetables. Englyst's concern that manufacturers may produce products high in RS, thus inflating dietary fiber values, appears to be exaggerated. Regulatory authorities can limit the RS content of manufactured foods by measuring it in the insoluble residue provided by the AOAC procedure. Gravimetric residues that don't have RS could certainly be obtained by incorporating a solubilization step with aqueous DMSO, to remove the RS from the residue of the AOAC, or any other gravimetric procedure (Brillouet et al. 1988; Englyst and Cummings 1988); however, the potential effect of DMSO on other than fiber fractions has not been thoroughly studied. Evidence is accumulating that starch, other than RS, is sometimes malabsorbed. However, this does not justify excluding the small RS fraction encountered from the dietary fiber concept (Englyst et al. 1983). All of the RS, as defined by Englyst et al. (1987), is included in the AOAC method.

Future simplification of the AOAC enzymatic-gravimetric method is desirable, but care must be given to preserve the advantages it presently has. Attempts to eliminate the need for ash and protein corrections as a means of simplification have not been successful as yet (Schweizer et al. 1984). Omitting proteolysis and replacing the heat-stable alpha-amylase (Termamyl) with an autoclaving step, thus making pH adjustments unnecessary (Li and Andrews 1988), have been suggested. However, autoclaving at uncontrolled pH might lead to severe fiber degradation and solubilization, resulting in fiber losses. Further, and most importantly, omitting the proteolysis step would increase the protein correction and make calculating the TDF (residue minus ash and protein) more critical and subject to error (especially in high-protein low-fiber foods). Other suggestions, namely to decrease test portion weights and/or reagent volumes and filtration times (hence decreasing costs) are worthwhile endeavors, but they should be validated through collaborative study before they are introduced into a well-standardized method (Lee 1991; Li and Andrews 1988).

Lee's Modification of the AOAC Method

Recently, at the 1990 American Association of Cereal Chemists (AACC) annual meeting, Lee presented the results of a collaborative study of a modification of AACC methods 32-05 (AOAC accepted method for TDF) and 32-21 (proposed AOAC method for SDF and IDF). This method differs from the currently accepted AOAC method in that (1) MES-TRIS buffer pH 8.2 replaces phosphate buffer pH 6.0, (2) a pH adjustment step for protease action is eliminated, (3) the volume of the reaction mixture is reduced, resulting in a reduction of total filtration volume, and (4) a pregelatinizing step and pH adjustment is eliminated. Lee reported that (1) the calculated TDF (by the sum of SDF + IDF) were in excellent agreement with the TDF values measured independently and (2) the overall precision for the SDF, IDF, and TDF determination was excellent: the mean RSD_r of SDF determination was 14.3% for cereal products and 12.2% for fruits and vegetables, and the mean RSD_R of IDF and TDF determination was equal to or less than 6.0% for both food groups.

In an Australian interlaboratory survey of fiber methods for cereal foods, also reported on at the 1990 AACC meeting, five foods were analyzed for fiber by 37 laboratories, utilizing their respective current methods (Mugford 1991). It was reported that (1) most of the laboratories selected the AOAC method for dietary fiber, (2) total dietary fiber precision in this study was satisfactory, and (3) should future food regulations require a 6% TDF for a label claim of "high in fiber," this study would conform to the requirement. The advantages and disadvantages of other fiber methods will not be discussed here because they were included in an earlier review (Asp and Johansson 1984). Two new gravimetric methods have recently been introduced and will be discussed in turn.

Mongeau and Brassard Method

The first of these methods is the Mongeau and Brassard (1986) method. The basis of this method is the measurement of NDF (giving the water insoluble fiber), to which is added the water-soluble fiber to give the TDF. In this method, insoluble dietary fiber is determined by the method of Goering and Van Soest (1970), using the hot weighing procedure. The modifications from the original method were as follows. Sodium sulfite was not added in the neutral detergent, to maximize the recovery of lignin (Robertson and Van Soest 1981). Fibertec extractors (Tecator, Sweden) were used instead of hot plates, and condensers and P2 glass filtering crucibles with a porosity of 40 to 90 um (Tecator, Sweden) were used. A rapid treatment with a freshly purchased mammalian alpha-amylase was performed on the fiber residue. Amylase solution was prepared by mixing 5 g alpha-amylase powder (Sigma Cat. #A6880) for 15 min with 100 ml buffer solution, pH 7 (61 ml 0.1M Na_2HPO_4 + 39 ml 0.1M NaH_2PO_4), centrifuging 10 minutes at 3000 rpm and filtering through a coarse sintered glass funnel or crucible with a porosity of 40 to 60 um. After refluxing with neutral detergent, the crucibles were filled with 10 ml cold alpha-amylase solution and 15 ml hot distilled water (the mixture being around 55°C). The crucibles were held for 5 minutes on the Fibertec heat extractor, suction was applied to the filter, and they were washed with hot distilled water. A No. 7 rubber stopper was used to seal the bottom of the crucibles. The crucibles were filled again with 10 ml cold alpha-amylase solution and 15 ml hot distilled water and were placed in an oven at 55°C. After 60 minutes of digestion, filtration was performed on the Fibertec cold extractor. The complete removal of starch was verified in food product residues, by adding a few drops of iodine solution. No blue color was generated after the crucibles were refrigerated. The residues were washed three times with hot distilled water and twice with acetone. After the acetone was evaporated, the crucibles were dried overnight at 100°C and weighed hot. The residue was ashed at 525°C for 4 hours, placed in an air-forced oven at 100°C overnight, and weighed hot. The insoluble fiber content of the sample was calculated as follows:

$$\%\ \text{insoluble fiber} = \frac{(\text{g fiber residue} - \text{g blank}) \times 100}{\text{g sample}}$$

Soluble fiber (the fiber material that dissolves in water at 100°C and precipitates in alcohol) was determined by alcohol precipitation, using the Fibertec-E (Tecator). A blank was run through the entire procedure. Approximately 0.5 g dry test portions were weighed in duplicate, to within 0.1 mg and placed in 50 ml, screw cap tubes. To gelatinize starch, 20 ml hot water (95 to 100°C) was added to each tube, and the tightly sealed tubes were heated in a boiling water-bath for 15 minutes from the onset of reboiling. The contents were mixed twice during the heating period. The tubes were cooled to 55°C; 2 ml amyloglucosidase solution (amyloglucosidase Sigma Cat. #A-9268) 15% v/v in 2.OM sodium acetate-acetic acid buffer, pH 4.5 were added, and the tubes were incubated at 55°C for 1.5 hours. The tubes were then placed in a boiling water-bath for 30 minutes, and mixed by hand at 5-minute intervals to extract the soluble fiber. Soluble fiber was separated by filtration into flasks on a Fibertec-E® filtration apparatus (Tecator, Höganäs, Sweden), using P3 crucibles, with a porosity of 10 to 40 um, containing glass-wool as a filtering aid. The tubes were rinsed with 10 ml hot water. After the crucibles were removed from the instrument, the tubing of the Fibertec-E was rinsed with 5 ml hot water, to avoid contamination between samples. Two ml of the amyloglucosidase solution were added to the supernatant in the incubation flasks and incubated at 55°C for 1.5 hours. Four volumes of 100% ethanol were added to the flasks, which were left covered at ambient temperature for 1 hour. The contents were filtered on a 50 ml medium Gooch crucible, with a porosity of 10 to 15 um, containing glass wool; the crucibles were rinsed twice with 75% ethanol and then with glass distilled acetone. The crucibles were dried overnight in a forced-air oven at 100°C and weighed hot. Fiber was ashed at 525°C for

4 hours, placed in a forced-air oven at 100°C overnight, and weighed hot. The soluble fiber content of the sample was calculated as follows:

$$\% \text{ soluble fiber} = \frac{(\text{g fiber residue} - \text{g blank}) \times 100}{\text{g sample}}$$

The TDF value was obtained by adding the soluble fiber value to that obtained with the NDF (neutral detergent fiber) procedure, modified as described above. The NDF + SOL fiber contains polysaccharides, lignin, some fiberbound protein, and other fiber-associated substances.

The hot weighing procedure is further described. Using an electronic analytical balance sensitive to 0.1 mg, in proximity to a forced-draft oven set at 100°C, the following procedure was followed: a hot crucible was placed on the balance plate for 1 minute to heat up the balance and then the balance was zeroed. At time 0, the crucible was removed from the oven and placed on the balance plate; time 20 sec: the weight of the crucible was recorded and the crucible was removed from the balance; time 30 sec: the zero of the balance was recorded and another crucible was placed on the balance plate. The deflection due to the temperature change of the balance was subtracted from each weight. Mongeau and Brassard (1979) described the hot weighing procedure (Goering and Van Soest 1970) for when a conventional balance is used. A comparison between the values obtained by the AOAC-TDF method and the NDF + SOL method showed the following results (Table 2-1).

The results showed that the NDF + SOL method was in agreement with the AOAC method ($r = 0.997$, $P<0.001$), in the laboratory of the author. Collaborative trials should bring out the robustness of the method and, should it pass the test, it will serve the community in providing an alternative gravimetric procedure. It is interesting to note that, for the substances studied, the NDF + SOL method, which was in agreement with the AOAC-TDF method, was also in agreement with the method of Englyst, using gas chromatography.

In a recently published paper, Mongeau and Brassard (1990b) showed excellent results when they carried out a collaborative study of a gravimetric method for determining insoluble and total dietary fiber. Total dietary fiber was calculated as the net residue of the soluble and insoluble fractions. Soluble fiber was precipitated using ethanol, after being treated with a heat stable amylase, amylogucosidase, and protease, remove protein and starch. For the insoluble fiber analysis test, protions were refluxed with neutral detergent and the residue was treated with pancreatic alpha-amylase, to remove residual starch. For the five foods tested, the Reproducibility Relative Standard Deviation (RSD_R) values for IDF ranged from 3.10 to 35.3%, for SDF from 10.40 to 68.29%, and for total dietary fiber the RSD_R values ranged from 4.13 to 17.94%. Using this

TABLE 2-1. Comparison of AOAC and NDF + Sol Methods for Total Dietary Fiber Values in Unprocessed and Processed Foods (% of Fresh Weight)

Sample	Moisture	AOAC TDF		NDF + SOL	
1. Wheat bran	9.7	45.48	45.45	44.06	44.19
2. Oat bran	3.2	17.09	17.09	18.81	19.97
3. Corn bran	8.4	63.81	63.22	63.58	63.38
4. Bran flakes	4.4	16.27	15.52	13.08	13.09
5. Corn flakes	2.7	3.50	3.46	0.75	0.82
6. Rice	10.9	1.02 ±	0.34[a]	1.29	1.10
7. Rice cereal	1.9	0.44	0.44	1.91	2.04
8. White flour	10.2	3.14	2.86	4.52	4.61
9. Broccoli	89.9	3.13	3.18	1.96	1.96
10. Cabbage	90.7	2.27	2.29	1.44	1.29
11. Turnip	88.3	2.79	2.80	1.76	1.72
12. Carrot	92.4	2.06	2.10	1.07	1.06
13. Baked beans	72.1	4.64	4.97	5.26	5.13
14. Soya beans	7.4	17.84 ±	0.25[a]	13.08	12.45
15. Raisins	7.5	3.19 ±	0.27[a]	3.60	3.33
16. Pectin powder	9.7	84.54 ±	1.28[a]	80.28	82.50

[a]Mean of 3-4 determinations ± SD

Reprinted with the permission of R. Mongeau and R. Brassard. The *Journal of Food Sciences*, 1986. 51:1334.

method and comparing it to the Englyst and AOAC method for 38 foods, Mongeau and Brassard (1990a) showed:

1. TDF values obtained with the method tested in the present study were in close agreement with those obtained using the Prosky TDF method (y = 1.03 × -0.13; r^2 = 0.98). The deviation of a y value from the regression line was within 0.23 g/100 g dietary fiber for 21 of the 25 non-cereal foods. The maximum deviation was -1.28 g/100 g dietary fiber for a finely ground breakfast cereal. However, this deviation mostly vanished when Celite was used as a filtering aid with both methods.
2. In low-lignin foods, the method agreed with the values found by Englyst. Mongeau could possibly be isolating the same component in both the NDF and soluble fractions because they analyze the NDF and soluble fractions on two different samples, thereby increasing the total through addition. This factor of possible overestimation should be looked into by the investigators.

The UED Method

Another enzymatic-gravimetric method, the urea enzymatic dialysis method (UED), has recently been proposed as an alternative to the

AOAC-TDF method (Jeraci et al. 1989). In this method, 8M urea hydrates and extracts starch and other water-soluble polysaccharides, a heat-stable amylase and dialysis digest and remove starch, and then protease is used to digest the plant proteins. The procedure involves the following.

Pretreat dialysis tubing in 10% ethanol-water. Remove dialysis tubing from 10% ethanol-water solution and close the end that was not inverted with clamp. Blow air gently into other end to open tubing. Attach inverted ends of four dialysis tubes to weighing station. Weigh 1 g test portions and transfer into each of the four dialysis tubes. Test portions weights may vary from 0.9000 to 1.0000 g. Remove tubing with test portion from weighing station, and add 30 mL 0.67% heat-stable amylase-8M urea solution to tubing. Unfold inverted end of tubing, close tubing with clamp, and gently swirl contents. If test portions are viscous, dialysis tubing may be gently squeezed to mix contents. Place dialysis tubing in beaker containing 8M urea for 3.5 to 4.5 hours at room temperature. Beaker (2L) will hold up to 30 tubes.

Remove tubing from beaker, open one end, and pipet 0.5 mL protease (Savinase) into tubing. Close tubing with clamp. Place tubing in water bath and hold for 2 to 28 hours at 50°C, with continuous change of water. Remove tubing from water bath, empty contents into 600 mL beaker, and rinse tubing into beaker with water.

Add 4 volumes of absolute ethanol to beaker, cover beaker, and hold at least 4 hours, to make filtration easier. Filter contents of beaker under vacuum through preweighed Whatman No. 54 paper (12.5 cm) on conical funnel; then wash beaker, paper, and residue with 80% ethanol. Place filter paper and residue in 105°C forced air over for 8 hours and then weigh. Analyze filter paper containing residue and filter paper blank in duplicate for protein, by using Kjeldahl method. If necessary, correct for nitrogen in filter paper. Multiply nitrogen result by 6.25 to determine protein.

Use duplicates for ash determination. Place filter paper containing residue and filter paper blanks in preweighed 50 mL beaker and incinerate 8 hours at 525°C. Remove beakers containing ash, place them in 105°C forced air oven for 1 hour, and then hot-weigh. Calculate TDF determined by UED from the formula described by Prosky et al. (1985).

$$\text{TDF, \%} = (\text{mg residue} - [(\text{\% protein in residue} + \text{\% ash in residue}) \times \text{mg residue}] - \text{blank} \times 100)/\text{wt sample, mg}$$

More details on how the procedure is set up, the reagents employed, and the method of hot weighing are described in detail in the paper. A comparison of results for dietary fiber using the AOAC and UED methods is recorded in Table 2-2.

The TDF by the AOAC enzymatic-gravimetric method was not significantly different from that determined by the UED method. The protein and ash in the AOAC residue were higher than those in the UED residues. Although resistant starch was negligible or absent in TDF residues of kale, using the UED method, as much as 3.5% resistant starch was found when using the AOAC method. The resistant starch of the other three products was green peas (0.4), broccoli (0.0), and baby lima beans (0.9). It was of interest to observe that no resistant starch was detected in the TDF, from mature bananas analyzed by the UED method. Englyst and Cummings (1986) reported that bananas contain up to 66% resistant starch, which is resistant to in-vitro digestion with pancreatic amylase. The method should be explored further and collaborated as an additional method for TDF analysis.

Berlin Method

Another gravimetric method is a modified version of that proposed by Thomas and Elchaly, in 1976, called the Berlin Method (Meuser et al. 1983; 1985; Becker et al. 1986). This method presents an analytical scheme for soluble and insoluble dietary fiber. The test sample is ground and defatted. It is then autoclaved, extracted, and centrifuged. The cen-

TABLE 2-2. Composition of Food Samples by AOAC Enzymatic-Gravimetric Method and Urea Enzymatic Dialysis Method (UED)[a]

	Total dietary fiber, %		Crude protein,[b] %		Ash,[c] %	
Sample	AOAC	UED	AOAC	UED	AOAC	UED
Sweet green peas	21.1	20.9	5.1	3.1	1.39	1.0
Broccoli	36.1	34.0	9.3	5.8	3.9	1.1
Baby lima beans	20.9	20.0	4.6	3.1	1.0	0.0
Green beans	30.0	30.0	4.0	3.6	6.1	2.7
Kale	35.5	32.5	11.1	6.9	9.5	2.3
Okra	41.8	40.6	10.1	5.2	6.6	2.7
Summer squash	21.4	20.2	4.5	3.3	3.1	0.8
Pectin	NA[d]	99.7	NA	2.3	NA	2.6

[a]Results corrected for water content. Crude protein determined by Kjeldahl analysis.
[b]Crude protein in AOAC residues significantly different than crude protein in UED residues ($P<0.05$).
[c]Ash in AOAC residues significantly different than ash in UED residues ($P<0.01$).
[d]Not analyzed. Reprinted with the permission of Jeraci et al. The *Journal of the Association of Official Analytical Chemists*. 1989. 72:679.

trifugate and extract are then separately hydrolyzed with amyloglucosidase and pancreation/trypsin. By altering the buffer combination and using a citrate/phosphate buffer, the amylolytic and proteolytic hydrolysis of the substrate was achieved, without having to introduce an intermediate separation step, minimizing a source of error. The enzymatic substrate hydrolysis was improved by using a magnetic stirrer that was on constantly. This improved the reproducibility of the results and also reduced the protein associated with the dietary fiber, from 7% to 4%. The soluble dietary fiber was determined after the micro- and macromolecular components were separated by ultrafiltration. Following enzymatic hydrolysis of the glucose polymers, proteins, and other macromolecular components, the substances in the retained fraction were determined gravimetrically as soluble dietary fiber. Figure 2-1 shows a schematic diagram of this method. Using the Berlin method, the dietary fiber content was determined in a number of breads, vegetables, legumes, and fruits, giving results comparable to the Asp method; however, the values for soluble dietary fiber were lower and those for insoluble dietary fiber were higher, in the Asp method.

Colorimetric Methods

Selvendran et al. (1989) have thoroughly discussed the colorimetric methods used to determine the sugars released on acid hydrolysis of dietary fiber. Southgate (1969), a pioneer in the dietary fiber field, was among the first to fractionate a food fiber into its components, soluble and insoluble noncellulosic polysaccharides (NCP), cellulose, and lignin. Dietary fiber was defined as nonstarch polysaccharides (NSP) (that is, the sum of NCP and cellulose). After extracting a food sample with 85% methanol, to remove free sugars, and acetone, to remove fat, a portion of the sample is treated with amylase overnight. The soluble dietary fiber fraction is then precipitated with 80% ethanol. The starch-free total dietary fiber residue is further fractionated into the soluble and insoluble fractions. Both fractions are hydrolyzed in 1N H_2SO_4, and the hexoses, pentoses, and uronic acids of each fraction are measured. The residue of insoluble fiber is extracted with 72% H_2SO_4, to remove and hydrolyze the cellulose and to determine the hexoses, pentoses, and uronic acids. The remaining residue is analyzed for lignin (Southgate 1981). The monosaccharides are measured as follows: hexoses by the anthrone and thiourea reagent, as described by Roe (1955); pentoses by the method of Mejbaum, as modified by Albaum and Umbreit (1947), with a slight modification (Southgate 1969); uronic acids by the carbazole method of Bitter and Muir (1962).

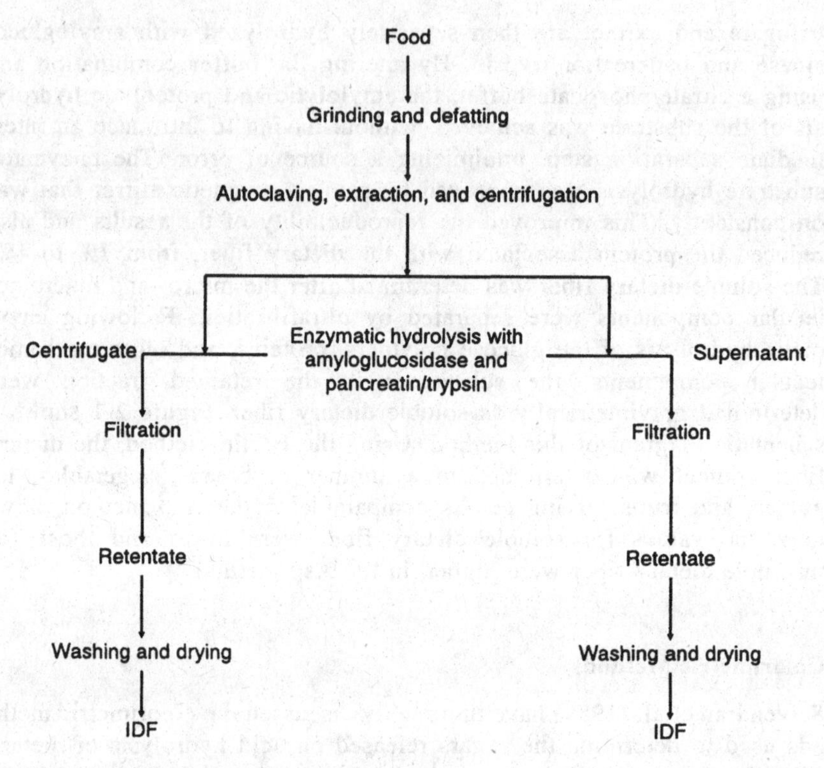

FIGURE 2-1. Schematics of the Berlin method.

Lignin, the residue insoluble in 72% H_2SO_4 was oxidized with permanganate, to obtain an estimate of heat-induced artifacts that remain insoluble after oxidation. Lignin is represented by the weight loss of the residue during this procedure.

The usual method for measuring uronic acid is that of Blumenkrantz and Asboe-Hansen (1973). In this method, dietary fiber is added to a dilute acid before it develops a chromophore with m-hydroxydiphenyl reagent, which contains sodium tetraborate in concentrated sulfuric acid, to give a product that is read at 520 nm in a spectrophotometer. There is very little interference from hexoses and pentoses. The modification introduced by Galambos (1967) significantly reduced the interference of hexoses. However, in the experience of many investigators, the ammonium sulfamate added in the carbazole (Dische 1947) method precipitates when the tubes are cooled.

Uronic acids can also be determined by the decarboxylation procedure used by Theander and Åman (1979), who found that the method of Bitter and Muir (1962) had low precision. Theander and Åman's method was rapid and direct and was based on the decarboxylation of the fiber fraction with hydroiodic acid. There was subsequent recording of the release of carbon dioxide trapped in a sodium hydroxide solution, by conductivity measurement (Bylund and Donetzhuber 1968). This method, unlike the colorimetric method, is unaffected by neutral carbohydrates. The molar yield of CO_2 has been shown to be on the order of 99% in the hydrochloric acid decarboxylation procedure of both D-glucuronic and D-galacturonic acids (Anderson and Garbutt 1961; Madson and Feather 1978). The studies of Theander and Åman (1979) showed that these two acids gave the same release of CO_2 by a more rapid hydroiodic acid decarboxylation procedure. It was found to be particularly useful for determining uronic acid in dietary fibers when D-galacturonic acid was present as a polymer. The hemicellulose uronic acid constituent, 4-0-methyl-D-glucuronic acid, gave similar results. Methyl ester linkages were found to be cleaved at an early stage, with strong acid treatment. A disadvantage of this method is that only one determination per hour can be performed with one piece of equipment.

Uppsala Method

In a recent communication (Theander and Westerlund 1986), the authors, in their third paper on studies on dietary fiber, came up with a method for improved procedures for analyzing dietary fiber. The paper deals with further developments for fractionation and GLC analysis of dietary fiber. The methods that resulted from this work—A, for water-soluble and insoluble dietary fiber—and the faster methods B and C, for total dietary fiber (Theander 1983; Theander and Westerlund 1986a, 1986b), are described below. Whole wheat flour, low-extraction wheat flour, and peeled raw and French-fried potatoes were analyzed in a collaborative study (Varo et al. 1983). Samples of oat, potato, rice, soy isolate, wheat bran (AACC-certified food grade), high extraction wheat flour, and a non-vegetarian and vegetarian mixed diet were also analyzed in an interlaboratory AOAC study (Prosky et al. 1984). Other samples that were analyzed were wheat bran of the cultivar Drabant, carrots of a local supermarket, wheat starch of the cultivar Folke, white wheat flour, and sugar beet fiber. A fractionation scheme for the three methods is shown in Figure 2-2.

In methods A and C, the total dietary fiber corresponds to the sum of neutral sugars and uronic acid constituents (all calculated as polysaccharides) and Klason lignin (Bethge et al. 1971), in the various fractions. Starch and its enzymatic hydrolysis products are removed. The contents

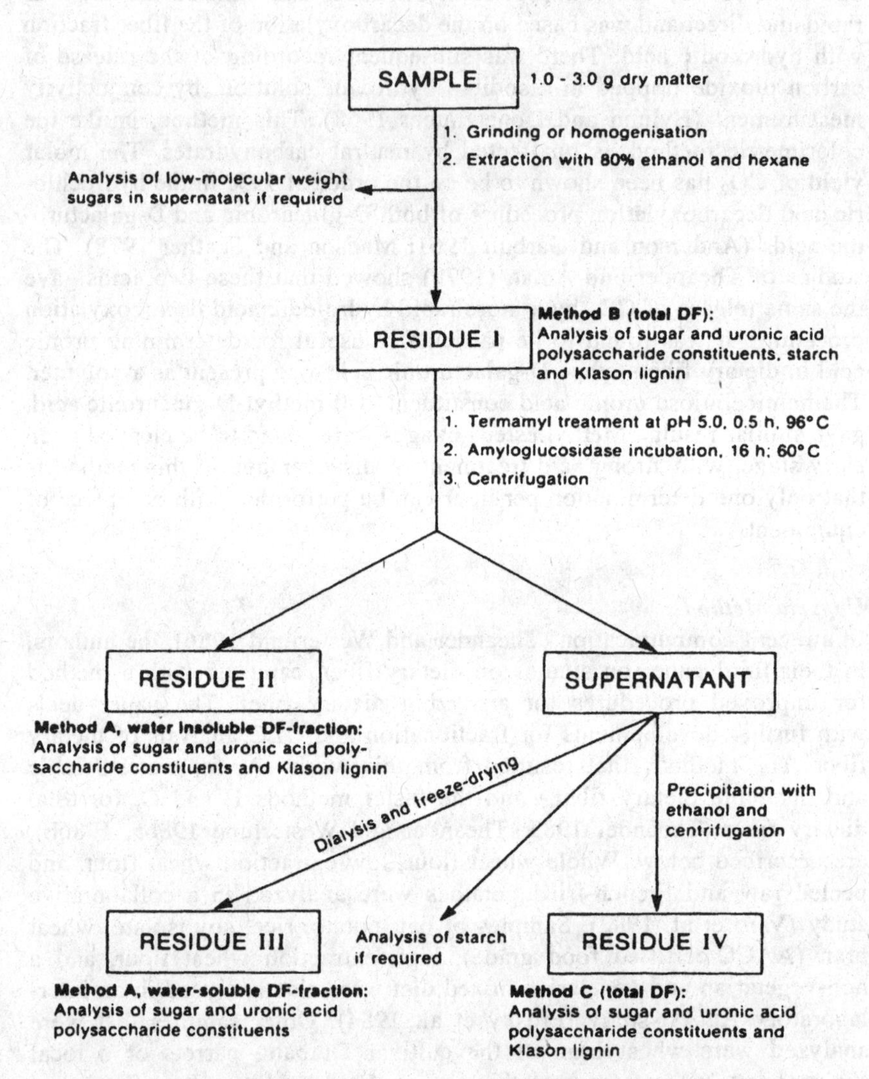

FIGURE 2-2. Fractionation scheme for methods A, B, and C for dietary fiber (DF). Reprinted from Theander, O., and E. Westerland. 1986b. Determination of Individual Components of Dietary Fiber. In *CRC Handbook of Dietary Fiber in Human Nutrition*, ed. G.A. Spiller. Boca Raton, FL: CRC Press Inc.

determined as pentose/deoxyhexose and hexoses/uronic acid were converted to polysaccharides by multiplication with the factors 0.88 and 0.90, respectively. In the most rapid method, B, the total dietary fiber is calculated as above, except that the starch value was subtracted from the dietary fiber glucan value obtained by GLC analysis. The results of the analysis are shown in Table 2-3. Comparing the results shows good agreement between A and C and also between B and C if the starch content is low. The authors recommend method A if separate analysis of water-insoluble and water-soluble dietary fiber components is required. The additional information that this method gives may not be warranted, as this method takes additional time. Method C is more suitable for analyzing starch-rich samples (for example, wheat flour) than method B, in which the necessary subtraction of a high starch value from a high total glucan value automatically lowers the precision of the dietary fiber determination. On the other hand, method B works well on products with low and moderate starch values. Method B correlated satisfactorily with the AOAC method and the enzymatic-gravimetric method of Asp et al. (1983), on samples used in the AOAC collaborative trial.

Since that time, Theander and Westerlund have developed a method for extracting soluble dietary fiber (Graham et al. 1988), which indicated that the yield and composition of the SDF fraction varied considerably with extraction condition and food type. So although extraction at high temperature gave, in general, the highest values for the SDF and the lowest for extraction in acid buffer, it would be more appropriate to use physiologically standardized conditions for the extraction of the SDF (Table 2-4). For example, in the case of potato and carrot, extraction following starch degradation at high temperature gave approximtely sixfold higher values than extraction with water at 38°C. The higher susceptibility of the fibers in those two foods to solubilization at high temperature led to a significantly increased solubility of fiber in water at 38°C, following pretreatment by boiling in ethanol. With the exception of potato and carrot, the values for SDF extracted by the two methods designed to prevent endogenous enzyme activity were very similar. The composition of the SDF was also affected by extraction conditions. For example, the glucose to arabinose ratios in the SDF in barley varied between 5.5 (method 4) and 11.3 (method 1), while the xylose to arabinose ratios remained around 1.4. The xylose to arabinose ratios in the SDF were about 1.5 for all cereals, regardless of the extraction method.

This study demonstrated that determining SDF is not dependent only on the extraction method employed, but also on the sample under analy-

Table 2-3. Comparison of Content of DF Components and Starch in Various Food Samples Using Methods A–C[a]

Sample	Anhydro Sugar Constituents										
	Rhamnose	Arabinose	Xylose	Mannose	Galactose	Glucose	Uronic Acids[c]	Klason Lignin	Total DF	Starch	Method
Raw potatoes	0.1	0.3	0.1	0.1	1.4	2.1	1.0	0.1	5.2	73.0	A
	0.1	0.3	0.2	0.1	1.6	1.8	1.0	0.4	5.5	71.6	C
French fried	0.1	0.3	0.1	0.1	1.5	3.9	1.0	0.8	7.8	69.7	A
potatoes	0.1	0.3	0.1	0.2	1.5	3.7	1.0	0.6	7.5	68.6	C
Low-extraction	trace	0.7	1.4	0.1	0.2	1.0	0.1	0.1	3.7	69.3	A
wheat flour	trace	0.8	1.3	0.2	0.2	0.9	0.2	0.2	3.8	69.7	C
Wheat bran	0.1	7.2	15.0	0.3	0.7	10.2	0.9	5.0	39.4	23.9	A
	0.1	7.5	15.0	0.5	0.7	9.2	0.9	4.6	38.5	23.6	C
Carrot 3	0.6	1.3	0.4	0.4	1.5	7.2	11.5	1.1	24.0	trace	A
	0.4	1.7	0.4	0.5	2.2	7.5	11.5	1.3	25.5	trace	C
Potatoes	0.2	0.3	0.2	0.2	1.7	5.2	1.3	0.4	9.5	66.7	B
(AOAC)	0.2	0.3	0.2	0.4	1.7	6.0	1.3	0.1	10.2	ND	C
Rice (AOAC)	trace	0.3	0.3	0.1	0.1	1.1	trace	0.7	2.6	77.3	B
	trace	0.2	0.2	0.3	0.1	1.5	trace	0.4	2.7	ND	C
Soya isolate	0.1	0.2	0.1	0.9	0.3	0.4	0.5	3.2	5.7	1.4	B
(AOAC)	trace	0.2	0.1	1.0	0.3	0.3	0.5	2.2	4.6	ND	C
Oats (AOAC)	trace	1.0	1.2	0.3	0.4	3.0	0.4	3.9	10.2	52.3	B
	trace	1.0	1.4	0.4	0.3	3.9	0.4	1.7	9.1	ND	C
Whole wheat	trace	2.4	3.7	0.3	0.6	0.6	0.6	2.2	10.4	59.5	B
flour (AOAC)	trace	2.2	3.5	0.4	0.3	2.6	0.6	1.2	10.8	ND	C
Vegetarian mixed	0.1	1.0	0.8	0.3	0.6	1.8	0.9	0.7	6.2	25.1	B
diet[b] (AOAC)	0.1	1.0	0.9	0.4	0.5	2.6	0.9	0.6	7.0	ND	C
Non-vegetarian mixed	0.1	0.9	0.8	0.4	0.8	2.0	0.8	1.1	6.9	31.7	B
diet[b] (AOAC)	0.1	0.8	0.8	0.5	0.5	2.5	0.8	0.7	6.7	ND	C

[a]Values expressed as percentages of dry matter. All samples analyzed at least in duplicate. [b]Defatted by ultrasonic extraction with hexane in method C. [c]Determined on original unextracted samples except for carrot, which had homogenized. Reprinted with permission of O. Theander and E. Westerlund, *Journal of Agricultural and Food Chemistry*, 1986, 34:334.

sis. One would also expect the use of dimethylsulfoxide, as in a recent paper of Englyst's, to increase the solubility of some polysaccharides.

These authors (Westerlund et al. 1990) have also arrived at a method for determining resistant starch (that is, starch not hydrolyzed by the combination of alpha-amylase and amyloglucosidase treatment).

The main advantage of this procedure, compared to those of Englyst et al. 1982, Englyst and Cummings 1985, and Siljeström and Asp 1985, is that the resistant starch is determined by a direct rather than a difference method and avoids the use of an unpleasant chemical, dimethylsulfoxide.

Method of Englyst and Cummings

The method for measuring dietary fiber as NSP, described by Englyst, has been reported earlier (Englyst et al. 1982; Englyst and Cummings 1984; Englyst 1985; Englyst and Hudson 1987). The method in its original form was rather complex, but the authors have improved it and simplified it, as a result of the demand for rapid routine analysis of dietary fiber. There has been no single paper available to describe how the method changes affect the individual neutral sugars and the resultant nonstarch polysaccharide/resistant starch values. These method changes, at least five in number, have included utilizing different enzymes, different pre- and post-treatment conditions, different acid hydrolysis steps, and others. The 1987 method utilizes 35°C for the primary acid hydrolysis step. The higher temperature of this step has been known to increase the formation of sulfate esters of neutral sugars, which resist hydrolysis during the secondary acid hydrolysis step and thus results in lower values for neutral sugars and NSP. In the later method, Englyst utilizes 25°C (ambient temperature) for the primary acid hydrolysis and a different condition for secondary acid hydrolysis, to give different values for neutral sugars.

Initially, Englyst and Cummings (1984) derivatized sugars by a technique in which the alditol acetates were prepared by using N-methyl-imidazole to catalyze acetylation before GLC. This step alleviates the need for the time consuming removal of borates by repeated evaporation with ethanol. To further simplify the method and allow for the measurement of NSP in laboratories without GLC, they recently introduced a modification by which the constituent sugars can be measured by colorimetry. One may measure total NSP or insoluble NSP by this method. Soluble NSP is calculated simply as the total NSP minus the insoluble NSP. The Englyst procedure for total dietary fiber is outlined below (Englyst 1989: Figure 2-3). When detailed information is not required, the colorimetric

TABLE 2-4. Content of Soluble Fiber Components Extracted under Four Different Conditions (m/g of Dry Matter)[a]

	Soluble nonstarch polysaccharide residue							
Extraction conditions	Rhamnose	Arabinose	Xylose	Mannose	Galactose	Glucose	Uronic acids	Total
Wheat								
Buffer (pH 5.0), 96°C, 60°C	0.3[a]	3.7[a]	5.8[a]	1.4[a]	3.4[ab]	9.6[a]	2.2[a]	26.5[a]
Water, 38°C	0.1[a]	2.6[b]	3.4[a]	0.9[b]	2.7[ac]	1.7[b]	0.8[b]	12.2[b]
Buffer (pH 1.5), 38°C	0.1[a]	2.7[b]	3.8[a]	0.7[b]	2.6[c]	1.3[b]	0.3[bc]	11.6[b]
Ethanol/water, 96/38°C	0.2[a]	2.5[b]	4.0[a]	1.4[a]	3.5[b]	2.1[b]	0.1	13.7[b]
SEM	0.03	0.2	0.3	0.1	0.1	0.4	0.1	0.9
Rye								
Buffer (pH 5.0), 95°C, 60°C	0.2[a]	9.6[a]	17.7[a]	2.7[a]	2.5[a]	7.2[a]	2.6[a]	42.4[a]
Water, 38°C	0.1[b]	7.2[b]	13.5[b]	1.1[b]	1.9[b]	2.5[b]	0.7[b]	27.1[b]
Buffer (pH 1.5), 38°C	0.1[b]	7.6[b]	13.9[b]	1.1[b]	1.9[b]	2.0[b]	0.6[b]	27.2[b]
Ethanol/water, 96/38°C	0.2[a]	5.8[c]	11.1[c]	1.7[c]	3.0[c]	2.6[b]	0.1[c]	24.4[b]
SEM	0.01	0.1	0.2	0.02	0.05	0.3	0.07	0.5
Barley								
Buffer (pH 5.0), 96°C, 60°C	0.3[a]	2.6[a]	3.4[a]	1.6[a]	2.0[a]	29.4[a]	3.1[a]	42.4[a]
Water, 38°C	0.2[a]	2.5[ab]	3.6[a]	0.9[a]	1.7[b]	23.2[b]	0.7[b]	32.7[b]
Buffer (pH 1.5), 38°C	0.2[a]	1.8[bc]	2.5[b]	0.8[c]	1.7[b]	11.2[c]	0.6[b]	18.7[c]
Ethanol/water, 96/38°C	0.2[a]	1.6	2.6[b]	1.3[d]	2.6[c]	8.8[c]	0.1[b]	17.1[c]
SEM	0.02	0.1	0.1	0.02	0.04	0.4	0.1	0.6
Oats								
Buffer (pH 5.0), 96°C, 60°C	0.3[a]	1.7[a]	2.0[a]	1.3[a]	2.6[ab]	25.7[a]	2.4[a]	36.0[a]
Water, 38°C	0.3[a]	1.6[a]	2.1[a]	0.9[b]	1.9[b]	36.3[b]	0.7[b]	43.7[a]
Buffer (pH 1.5), 38°C	0.3[a]	1.5[ab]	1.8[ab]	0.6[c]	1.9[b]	19.7[c]	0.8[b]	26.6[b]
Ethanol/water, 96/38°C	0.3[a]	1.3[b]	1.6[ab]	1.2[ab]	2.9[a]	16.1[c]	2.6[a]	26.0[b]
SEM	0.02	0.1	0.1	0.05	0.1	1.3	0.1	1.4
Potato								
Buffer (pH 5.0), 96°C, 60°C	1.0[a]	1.9[a]	0.6[a]	1.4[a]	9.6[a]	2.7[a]	9.6[a]	26.9[a]

TABLE 2-4. Continued

Extraction conditions	Soluble nonstarch polysaccharide residue							
	Rhamnose	Arabinose	Xylose	Mannose	Galactose	Glucose	Uronic acids	Total
Water, 38°C	0.1^{b}	0.4^{b}	0.4^{b}	0.7^{b}	1.6^{b}	0.7^{b}	1.6^{b}	5.5^{b}
Buffer (pH 1.5), 38°C	0.1^{b}	0.5^{b}	0.4^{b}	0.6^{b}	1.7^{b}	0.8^{b}	1.1^{b}	5.2^{b}
Ethanol/water, 96/38°C	0.3^{c}	0.4^{b}	0.6^{a}	0.7^{b}	2.7^{b}	1.2^{b}	3.2^{c}	9.0^{c}
SEM	0.02	0.03	0.02	0.03	0.2	0.2	0.1	0.3
Carrot								
Buffer (pH 5.0), 96°C, 60°C	4.2^{a}	11.3^{a}	0.7^{ab}	1.3^{a}	21.4^{a}	2.2^{b}	60.9^{a}	101.9^{a}
Water, 38°C	0.3^{b}	1.9^{b}	0.5^{b}	0.9^{b}	5.5^{b}	1.1^{b}	4.2^{b}	14.4^{b}
Buffer (pH 1.5), 38°C	0.3^{b}	2.8^{b}	0.6^{b}	1.0^{b}	5.3^{b}	1.2^{b}	3.4^{b}	14.5^{b}
Ethanol/water, 96/38°C	0.8^{b}	2.1^{b}	0.8^{b}	1.0^{b}	6.5^{b}	1.6^{ab}	20.9^{c}	33.7^{c}
SEM	0.1	0.2	0.03	0.04	0.4	0.1	1.4	1.7
Lettuce								
Buffer (pH 5.0), 96°C, 60°C	1.4^{a}	3.8^{a}	0.7^{ab}	1.0^{a}	8.6^{a}	1.5^{a}	32.1^{a}	49.2^{a}
Water, 38°C	0.2^{b}	3.1^{b}	0.5^{c}	1.2a	8.4^{a}	1.5^{a}	6.8^{bc}	21.7^{b}
Buffer (pH 1.5), 38°C	0.2^{b}	2.6^{c}	0.6^{bc}	1.2^{a}	6.9^{b}	1.3^{a}	6.0^{b}	18.8^{b}
Ethanol/water, 96/38°C	0.5^{b}	1.6^{d}	0.8^{a}	1.0^{a}	4.7^{c}	1.6a	8.8^{c}	19.1^{b}
SEM	0.05	0.1	0.02	0.04	0.1	0.1	0.3	0.5
Pea								
Buffer (pH 5.0), 96°C, 60°C	0.6^{a}	5.1^{a}	1.4^{a}	1.6^{a}	4.4^{a}	2.7^{a}	10.5^{a}	26.3^{a}
Water, 38°C	0.3^{b}	2.2^{b}	0.8^{b}	0.8^{b}	2.5^{b}	1.0^{b}	4.2	11.7^{b}
Buffer (pH 1.5), 38°C	0.2^{b}	1.5^{b}	0.6^{b}	0.6^{b}	1.9^{b}	0.7^{b}	2.0^{c}	7.5^{c}
Ethanol/water, 96/38°C	0.3^{b}	1.7^{b}	0.9^{b}	0.9^{b}	2.5^{b}	1.2^{b}	3.3^{b}	10.7^{b}
SEM	0.04	0.2	0.04	0.06	0.1	0.1	0.2	0.4

[a]Means and pooled standard error of means (SEM) of triplicate analyses. For any sample, values within a column that do not share a common superscript differ significantly ($P<0.001$).
Reprinted with permission of H. Graham et al., *Journal of Agricultural and Food Chemistry*, 1988, 36:496.

method is quite suitable for determining total, soluble, and insoluble nonstarch polysaccharide (NSP).

In the original procedure (Englyst 1981), starch was removed from the gelatinized food sample, using amyloglucosidase for 16 hours at 45°C. In the second paper (Englyst et al. 1982), using pancreatic alpha-amylase and pullulanase for 16 hours at 42°C, instead of amyloglucosidase, was recommended. The procedure was later modified to incorporate a DMSO

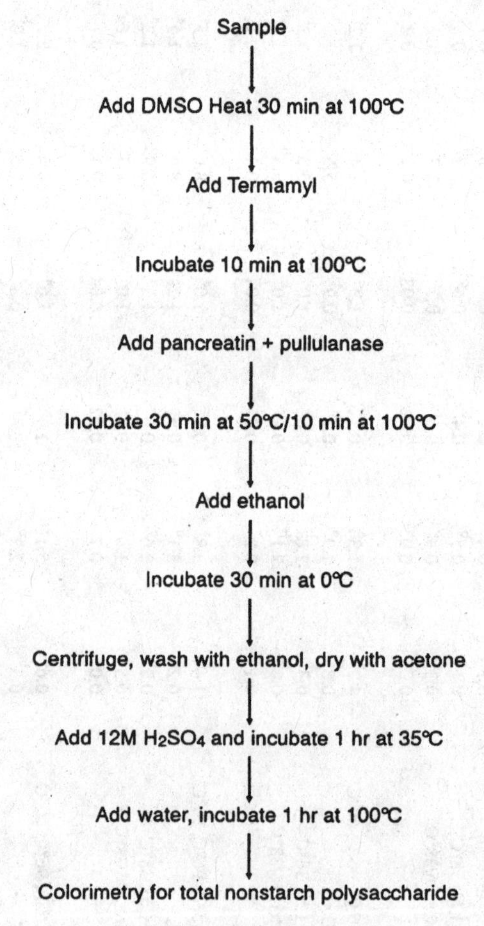

FIGURE 2-3. Schematics of colormetric procedure for determining nonstarch polysaccharides. For measuring insoluble dietary fiber, replace the ethanol with pH 7 buffer and extract for 0.5 hr at 100°C.

step at 100°C for 1 hour, in order to solubilize the resistant starch (Englyst and Cummings 1984).

In a paper describing the practical problems with the GC measurement of the neutral sugars as alditol acetates, Cooper et al. (1989) found that the capillary columns in the GC were found to deteriorate rapidly after three or four sample injections, thus making the Englyst determination very costly and unreliable. It was thought that the residual alkaline water present in the derivative builds up on the active stationary phase of the GC column. Cooper et al. first modified the system by using an empty silica pre-column placed between the injector and the main analytical column. Secondly, she prepared an alditonitrile acetate derivative, instead of an alditol acetate derivative. The former procedure is anhydrous, and any traces of water would be consumed during the preparation by the conversion of acetic anhydride to acetic acid. The results showed that, by using the pre-column and the alternative derivatization, up to 20 samples could be analyzed before the column deteriorated.

Although the alditonitrile acetate derivatives took longer to prepare, the running times on the GC were reduced from 22 to 12 minutes, since a shorter analytical column could be used.

One of the unanswered questions regarding the currently available enzymatic/gravimetric methods, which are followed by instrumental analysis (i.e., those methods that ultimately break down the fiber structure and analyze for monomeric components), for the analysis of dietary fiber, is whether the hydrolysis steps of the method release all the component saccharides from the isolated fiber residue obtained from the enzymatic digestion process. For oat groats (groats, oatmeal, or porridge oats), Frølich and Nyman (1988), using the Theander method, report a soluble fiber neutral sugars content of 2.9%, a total fiber neutral sugars content of 8.5%, and a lignin content of 2.0%. Shinnick et al. (1988), also using the Theander method, report a soluble fiber neutral sugars content of 4.8%, a total fiber neutral sugars content of 8.5%, and a lignin content of 3.3%. Cummings and Englyst (1987), using a method based on a modification of the Southgate method, report soluble fiber neutral sugars of 4.7% and 3.9% and total fiber neutral sugars of 7.6% and 6.9% for two samples, respectively. No lignin values are reported by these authors. In comparison to the results obtained by the above authors, Anderson and Bridges (1988), also using what they describe as a modification of the Southgate procedure, report a soluble fiber neutral sugars level of 5.3%, a total dietary fibers neutral sugars level of 9.1%, and a lignin value of only 1.0%. The main difference between the Anderson procedure and the others is the dissolution of the soluble fraction from the total dietary fiber residue, with water (100°C), filtration, and drying of the soluble fraction at 45°C, under a filtered air stream. A possible explanation for the increased neutral sugars levels, particularly the soluble fraction neutral sugars and the decreased lignin level in the Anderson results, could be that all the neutral sugars are not completely separated from the lignin in the analytical procedures used in the other laboratories or during the hydrolysis procedure per se used in the Anderson laboratory. The neutral sugars may actually be hydrolyzed in the dissolution and isolation steps carried out in the Anderson laboratories, prior to the actual hydrolysis step itself.

Similar effects have been seen for oat bran. Frølich and Nymann (1988), using the Theander method, report a soluble fiber neutral sugars content of 5.6%, a total fiber neutral sugars content of 11.5%, and a lignin content of 3.0%, for coarse ground bran. For fine ground bran, they report 5.2%, 12.1%, and 3.0%, respectively. Shinnick et al. (1988), also using the Theander method, report a soluble fiber neutral sugars content of 6.9%, a total fiber neutral sugars content of 14.1%, and a lignin content of 3.8%. On the other hand, Anderson and Bridges (1988), using the modification of the Southgate procedure, report a soluble fiber neutral sugars level of 7.7%, a total dietary fibers neutral sugars level of 14.0%, and a lignin value of only 1.6%. As before, Anderson and Bridges resolubilize the soluble dietary fiber fraction in hot water and isolate by drying with warm air. Granted that the authors reporting these results are not all using identical samples for their analyses, but the trend exists and the differences in quantities obtained are sufficient to suggest that the extra solubilization step of Anderson and Bridges appears to release neutral sugars associated with the lignin. The neutral sugars are thus solubilized, transferring to the soluble fiber fraction, whereas the other methods simply may not be completely hydrolyzing the neutral sugars associated with the lignin.

While the above authors did not necessarily use the same oat groat or bran samples, Lopez-Guisa et al. (1988) analyzed bleached and unbleached oat hulls from the same source, using the Theander method. The authors do not indicate if the actual samples analyzed were from the same lot prior to processing and bleaching. They do, however, report that the soluble fiber neutral sugars level increases from 0.5% to 3.7%, the total fiber neutral sugars level increases from 60.9% to 72.8%, and the lignin content dropped from 17.5% to 11.4%, when comparing the heat processed oat hulls to the bleached oat hulls. Heat processing consisted of heating at 91°C to 121°C, for 1 to 3 minutes. Bleaching was done in dilute hydrogen peroxide, at pH 10 to 12, 50 to 100°C for 1 to 2 hours. If all the neutral sugars present in the fiber fractions were released from the lignin during the analytical procedure, similar values for neutral sugars should have been obtained for both the heat processed and bleached samples. Apparently, the bleaching process is effectively hydrolyzing some of the saccharides bound to the lignin.

It would appear that some additional research may be necessary on the enzymatic/gravimetric/instrumental methods, to assure that all the neutral sugars are released hydrolytically and subsequently quantitated in their proper respective fractions during the course of the analytical procedures.

Berry makes a strong case for including resistant starch as a component of dietary fiber (Berry 1985). In Table 2-5, one can see the effect of omitting resistant starch. Englyst's values for dietary fiber (NSP) are probably lower than those reported by Southgate because of the efficient removal of starch and a decision to dispense with the estimation of lignin. This table shows that only cornflakes has significant resistant starch—approximately 3%. If resistant starch becomes a problem, it can be determined in any of the other methods by using a DMSO extraction step, although caution may be in order, to study the effects of DMSO on other fiber fractions. Termamyl removes all starch except for resistant starch, if samples for the AOAC method are correctly prepared for analysis. Resistant starch is also undigested in vivo, since it is totally recovered from ileostomy effluents. It is, however, fermented in the colon, as are other fiber polysaccharides. Therefore, it should be included in dietary fiber.

TABLE 2-5. Relative Levels of Resistant Starch and Nonstarch Polysaccharide in a Range of Processed Foods.*

	g/100g dry matter			
Food	Starch	RS	Total NSP	RS: Total NSP
White bread	77.5	1.15	2.63	1 : 2.3
Brown bread	67.7	0.91	7.25	1 : 8.0
Wholemeat bread	64.6	0.80	9.58	1 : 12.0
All Bran (Kelloggs)	29.1	0.15	23.68	1 : 158.0
Weetabix	62.3	0.26	10.41	1 : 40.0
Shredded wheat	64.4	0.88	10.73	1 : 12.2
Cornflakes	78.9	3.11	0.68	4.6 : 1.0
Rice Krispies	73.9	0.24	0.89	1 : 3.7
Digestive wheatmeal biscuits	46.4	0.13	2.97	1 : 22.9
Rich tea biscuits	52.3	0.19	2.22	1 : 11.7

*Data shown are obtained from Englyst et al., 1983, *Journal of the Science of Food and Agriculture*, 34:1434-1440.

Englyst and Cummings give two reasons for not including resistant starch as dietary fiber (a third would be, of course, to favor their method):

- Food manufacturers might produce foods with high resistant starch contents, to inflate fiber values.
- Fiber contents of foods would vary with processing, thus making aggregation of dietary fiber values for food composition tables more difficult.

Neither argument is pertinent. If government food authorities do not want foods to contain more resistant starch, they have means other than a fiber method to avoid it. Nobody uses urea to inflate protein contents of foods, although this would work with respect to the Kjeldahl method. Besides, it is not known whether, for example, microcrystalline cellulose is a better fiber than resistant starch, and resistant starch can be determined in gravimetric residues obtained in the methods of Asp or AOAC, as noted above. Processing, including cooking at home, changes the contents of many nutrients (for example, vitamins) including fiber. Resistant starch is only one of several fiber components that can change.

It could also be argued that, in the context of a method, resistant starch and lignin, if excluded from fiber, would inflate available carbohydrates. More importantly, however, Englyst and Cummings state that in most foods lignin is quantitatively unimportant and that "the present small amount of resistant starch is nutritionally insignificant." Why, then, worry so much about it?

To quote Berry, "however, it is an inescapable fact that people do not eat raw flour, raw potato or, for that matter, any other very starchy foods in their raw state. Most of the food that is eaten in the United Kingdom is processed in some way, either in the home or by commercial food processors. Starch is the major carbohydrate consumed in this country, and its nutritional value as a complex carbohydrate has been widely acknowledged in recent years. Removal of its present status as potential man-made dietary fibre can only be described, appropriately perhaps, as a retrograde step."

In addition, the Englyst procedure gives low values of dietary fiber, even for foods with negligible resistant starch and lignin. It is very likely that such low values are due to destructive losses of sugars during hydrolysis and/or to incomplete hydrolysis of polysaccharides to monosaccharides and uronic acids. The Englyst procedure does not account for these losses because the internal standard is added after hydrolysis only. Instead, a uniform 10% loss is assumed, but it is well known that different sugars are destroyed at different rates. Interestingly, the method of Theander, which is also based on hydrolysis of polysaccharides to the constituent sugars and GLC, gives values in agreement with AOAC and Asp, in part because corrections for hydrolysis losses are made.

Selvendran et al. (1989) reported that he carried out the Englyst hydrolysis procedure, that is, by the modified Saeman method (Saeman et al. 1963). In this procedure, the dietary fiber preparation was dispersed in 12M (72% w/w) H_2SO_4 and left for 1 hour at 35°C, to solubilize all polysaccharides. The mixture was diluted with distilled water and hydrolyzed for 2 hours with continuous mixing in a boiling water bath (Method A). He also hydrolyzed with M H_2SO_4 for 2 hours at 100°C (Method B), to test the sugar released from a variety of cereal-based products. The sugars released were analyzed as alditol acetates. In all cases it appears that (1) the amount of xylose released upon hydrolysis by method B was significantly greater, sometimes as much as 30%-40% greater than by method A; (2) the amount of arabinose released was 10%-15% greater using method B; and (3) galactose release was comparable, using the two methods of analysis. The higher amounts of xylose and arabinose released by method B may indicate that the arabinoxylans from the dietary fiber of endospermous tissues are readily solubilized and partially degraded by method A, whereas those from the dietary fiber of lignified tissues are released more slowly, with correspondingly small losses by degradation. The values of those sugars (xylose, arabinose, and galactose) recovered from dietary fiber preparation by method A, expressed as a percentage of method B, is given in Table 2-6.

These observations were verified, using dietary fiber preparations from wheat bran and wheat flour. Losses in pentoses, notably xylose, were observed in these two cereal products following 12M H_2SO_4 hydrolysis at 35°C. This point should be considered when using the En-

TABLE 2-6. Sugars Recovered from DF Preparations by Method A, Expressed as a Percentage of Method B.

Cereal products	Xylose	Arabinose	Galactose
All Bran (Kelloggs, U.K.)	72	83	90
"Farm House" bran flakes	67	78	100
Bran Flakes (Kelloggs, U.K.)	78	91	95
Shredded wheat	61	74	97
"Wheetabix" wheat biscuit	81	90	100
Round puffed wheat	82	89	94
Special K (Kelloggs, U.K.)	64	73	82
Oat-based cereal	73	85	100

Reprinted from Selvendran, R. et al. *Modern Methods of Plant Analysis*, Volume 10, Plant Fibers, 1989:253.

glyst procedure. Selvendran recommends treatment with 12M H_2SO_4 for 2 to 3 hours at 20°C, followed by dilution to M strength and hydrolysis at 100°C for 2.5 hours. The sulfuric acid may be removed by treatment with 0.25M $Ba(OH)_2$ (Selvendran et al. 1979) or with solid $BaCO_3$ (Englyst et al. 1982).

There are other doubts, though, about the validity of the procedure used to hydrolyze the dietary fiber preparations. Attention was recently drawn to the fact that, when enzyme digests of starch-rich foods are treated with ethanol in order to precipitate dietary fiber-polymers, significant amounts of glucose and other low molecular weight oligosaccharides are co-precipitated with the dietary fiber polymers. The co-precipitated glucose and oligosaccharides tend to inflate values for the apparent glucan content of these preparations, but can be removed by dialysis (Marlett 1989). In a recently conducted AOAC collaborative study, sugars were removed by decantation from an 85% methanol solution because the residues of some fruits contained large amounts of sugar, making the residues sticky (Prosky et al. 1991) and very hard to dry. This may be a problem with the Englyst procedure.

Method of Faulks and Timm

Faulks and Timm (1985) have adopted and incorporated a number of ideas from the methods of Theander and Åman (1979) and Englyst and Cummings (1984), for dietary fiber analysis when removing soluble and retrograded starch. The procedure is outlined in Figure 2-4. Most of the starch is removed from the gelatinized material, using Termamyl at 100°C for 15 minutes.

Absolute ethanol is added to the cooled mixture, to a concentration of 80% ethanol, and the precipitate that forms is removed by centrifugation. The precipitate is then dispersed in DMSO at 100°C and incubated with amyloglucosidase at 37°C for 15 minutes. The DMSO solubilizes resistant starch and residual dextrins, which are both hydrolyzed. The extract is cooled, made up to 80% ethanol with absolute ethanol, centrifuged, and the precipitate dried in an oven at 100°C. Selvendran and King (unpublished results) have found that freeze-drying the precipitate makes it more dispersable in dilute acids for hydrolysis prior to sugar analysis, as well as for methylation analysis. Selvendran et al. (1979) used two procedures for the hydrolysis of dietary fiber to their constituent sugars (1) M H_2SO_4 hydrolysis for 2.5 hours at 100°C and (23) treatment with 72% H_2SO_4(12M) for 3 hours, followed by dilution to M acid and hydrolysis for 2.5 hours at 100°C. They also treated dietary fiber preparations with 72% H_2SO_4 at 20°C for 2 hours prior to post-hydrolysis,

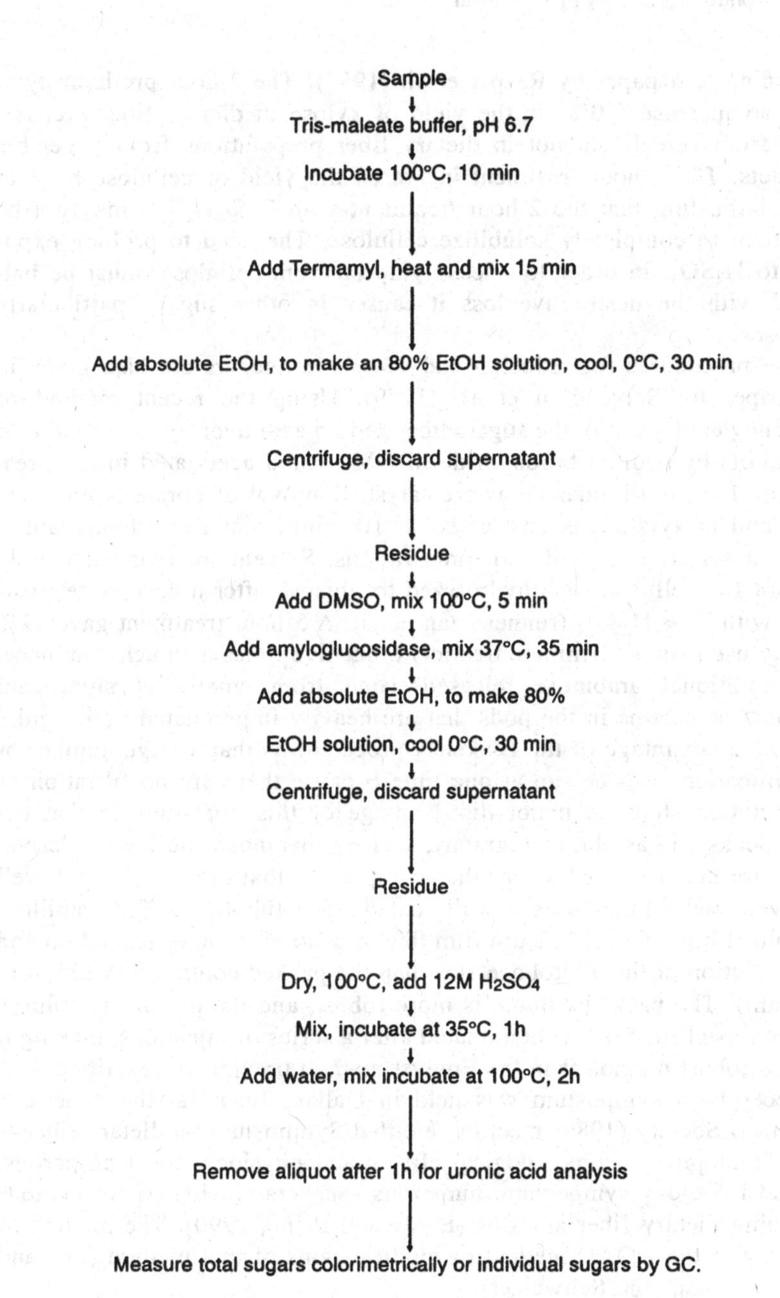

FIGURE 2-4. Procedure of Faulks and Timm's for determining dietary fiber.

according to a paper by Rasper et al. (1981). The 2-hour pre-hydrolysis gave an increase (10%) in the yield of xylose in dietary fiber preparations from cereal, but not in dietary fiber preparations from vegetable products. The 3-hour treatment increased the yield of cellulose by 8 to 10%, indicating that the 2-hour treatment with 72% H_2SO_4 may not be sufficient to completely solubilize cellulose. The need to prolong exposure to H_2SO_4, in order to release glucose from cellulose, must be balanced with the destructive loss it causes in other sugars, particularly pentoses.

The procedures for reducing and acetylating sugars are also given in the paper by Selvendran et al. (1979). Using the recent method of Blakeney et al. (1983), the sugars liberated on acid hydrolysis are reduced to alditols by sodium borohydride in DMSO and acetylated in the presence of 1-methyl-imidazole as a catalyst. Removal of borate is unnecessary, and acetylation is completed in 10 minutes at room temperature. Using a selection of cell wall preparations, Selvendran found that only 65% of the cellulose was hydrolyzed to glucose after a 45-minute treatment with 72% H_2SO_4 (runner bean pods). A 3-hour treatment gave 12% less xylose than a 45-minute treatment, but five times as much arabinose. The additional arabinose released came from small but significant amounts of pectins in the pods that are heavily impregnated with lignin. The main advantage of the Blakeney procedure is that a large number of determinations can be run at one time because there are no filtration or precipitation steps. A major disadvantage of this procedure is that the early peaks on gas-chromatography, such as rhamnose and fucose derivatives, are contaminated with other compounds that are usually not well resolved. Selvendran uses a wall-coated open tubular OV225 capillary column, 15 m × 0.5 mm, 1 μm film thickness to give better resolution and faster elution of the alditol acetates than the packed column (OV225, 4 m × 3 mm). The packed column is more robust, and the top of the column can be repacked if it is contaminated after a series of injections, making it a more robust method than the Englyst method previously described.

Recently, a symposium was held in Dallas, Texas, at the American Chemical Society (1989) meeting, entitled Symposium on dietary fiber—new developments: physiological effects and physiochemical properties. At that three-day symposium, numerous speakers elucidated on methods for doing dietary fiber analysis (Furda and Brine, 1990). The method of choice was the AOAC method or modifications of that method (Lee and Hicks, Li, Asp, and Schweizer).

In his paper reviewing the total dietary methodology, Hall (1989) stated that crude fiber, acid detergent fiber, neutral detergent fiber, the Southgate approach, Englyst modifications, and the Theander and Westerlunds analytical scheme gave adequate values and defined some of the constituents of fiber. However, the methodology is complex and time consuming and is not a cost-effective way to accumulate data bases for statutory labeling requirements. Hall called for the development of a rapid regulatory method, in addition to the chemical research method already named. The AOAC method was developed with this in mind and was subsequently adopted as the official method for total dietary fiber in the United States. It closely mimics the action of the upper G.I. tract, by using proteolytic and amylolytic enzymes and yields results that are comparable with the physiological definition of total dietary fiber. This method is also listed as the AACC Method 32.05 (1988) and has some limitations. It does not recover some of the indigestible polysaccharides that are currently being added to food such as polydextrose (Furda 1989).

In a recent communication from P.J. Wagstaffe of the Commission of the European Communities, Community Bureau of Reference (BCR), we were informed that the BCR program had completed several studies that have led to the establishment of certified values for dietary fiber measured by the AOAC method for three food reference materials. The reference materials are haricot beans, wheat flour, and rye flour. The AOAC method was chosen over the Englyst procedure because of their lower RSD_R values. For haricot beans the RSD_R was 3.8% using the AOAC method and 11.0% using the Englyst procedure. For wheat flour and rye flour the RSD_R's using the AOAC method were 10.1% and 4.5%, respectively, and using the Englyst procedure were 29% and 11%, respectively.

The remaining part of the methodological development is to adapt the method for total dietary fiber, to be applied to a method for soluble and insoluble dietary fiber. This has been partially accomplished (Prosky et al. 1990).

References

Albaum, H.G., and W.W. Umbreit. 1947. Differentiation between ribose-3-phosphate and ribose-5-phosphate by means of the orcinol pentose reaction. *Journal of Biological Chemistry* 167:369–376.

American Association of Cereal Chemists, AACC. 1978. *Approved methods of the AACC, Revisions: Method 32-20* AACC, Saint Paul, MN.

American Association of Cereal Chemists, AACC. 1988. *Approved methods of AACC: Method* 32.05 AACC, Saint Paul, MN.

Anderson, D., and S. Garbutt. 1961. Studies on uronic acids. III. An investigation, using ^{14}C compounds, of acid decarboxylation time. *Talanta* 8:605–611.

Anderson, J.W. 1982. Dietary fiber and diabetes. In *Dietary Fiber in Health and*

Disease. ed. G.V. Vahouny and D. Kritchevsky. pp. 151-167, New York: Plenum Press.

Anderson, J.W., and S.R. Bridges. 1988. Dietary fiber content of selected foods. *American Journal of Clinical Nutrition*. 47:440-447.

Asp, N.-G., C.-G. Johansson, H. Hallmer, and M. Siljeström. 1983. Rapid enzymatic assay of insoluble and soluble dietary fiber. *Journal of Agricultural and Food Chemistry* 31:476-482.

Asp, N.-G, and C.-G. Johansson. 1984. Dietary fibre analysis. *Nutrition Abstracts and Reviews in Clinical Nutrition* 54:735-752.

Asp, N.-G., I. Furda, J.W. DeVries, T.F. Schweizer, and L. Prosky. 1988. Dietary fiber definition and analysis. American Journal of Clinical Nutrition 47:688-689.

Baird, I.M., and M.H. Ornstein. eds. 1981. *Dietary Fibre: Progress Toward the Future*. Manchester, Kellogg Company of Great Britain.

Becker, H.G., W. Steller, W. Feldheim, E. Wisker, W. Kulikowski, P. Suckow, F. Meuser, and W. Seibel. 1986. Dietary fiber and bread: intake, enrichment, determination, and influence in colonic function. Cereal Foods World 31:306-310.

Belo, P.S., and B.O. de Lumen. 1981. Pectic substance content of detergent-extracted dietary fibers. *Journal of Agricultural and Food Chemistry* 29:370-373.

Berry, C. 1985. Putting a figure on dietary fibre. *Nutrition and Food Sciences* March/April:8-10.

Bethge, P.O., R. Radestrom, and O. Theander. 1971. *Communication from the Swedish Forest Products Laboratory* 63B:1-50, Stockholm, Sweden.

Bitter, T., and H.M. Muir. 1962. A modified uronic acid carbazole reaction. *Analytical Biochemistry* 4:330-334.

Blakeney, A.B., P.J. Harris, R.J. Henry, and B.A. Stone. 1983. A simple and rapid preparation of alditol acetates for monosaccharide analysis. *Carbohydrate Research* 113:291-299.

Blumenkrantz, N., and G. Asboe-Hansen. 1973. New method for quantitative determination of uronic acids. *Analytical Biochemistry* 54:484-489.

Brillouet, J.-M., X. Rovau, C. Hoebler, J.-L. Barry, B. Carre, and E. Lorta. 1988. A new method for the determination of insoluble cell walls and soluble nonstarchy polysaccharides from plant materials. *Journal of Agricultural and Food Chemistry* 36:969-979.

Browne, C.A. 1940. The origin and application of the term nitrogen-free extract in the valuation of feeding stuffs. *Journal of the Association of Official Agricultural Chemists* 23:102-108.

Burkitt, D.F., A.R.P. Walker, and N.S. Painter. 1972. Effect of dietary fibre on stools and transit-times, and its role in the causation of disease. *Lancet* 2:1408-1412.

Bylund, M., and A. Donetzhuber. 1968. Semi-micro determination of uronic acids. *Svensk Papperstidning* 15:505-508.

Cleave, T.L., and G.D. Campbell. 1966. *Diabetes, coronary thrombosis and the saccharine disease*. Bristol: John Wright.

Cooper, S.J., R.C. Palmer, and D.C. Porter. 1989. Investigation of practical problems with the GC measurement of neutral sugars in the modified Englyst dietary fibre method. *Research report 65. The British Food Manufacturing Industries Research Association*. Surrey, England.

Cummings, J.H., and H.N. Englyst. 1987. The development of methods for the measurement of 'dietary fibre' in food. In *Cereals in a European Context—First European Conference on Food Science and Technology*, ed. I.D. Morton. pp. 188-220 New York: VCH.

Dietary Fiber, Basic and Clinical Aspects. 1986. ed. G.V. Vahouny and D. Kritchevsky New York and London: Plenum Press.

Dietary Fiber-New Developments. 1990. ed. I. Furda and C. Brine. New York and London: Plenum Press.

Dische, Z. 1947. A new specific color reaction of hexuronic acids. *Journal of Biological Chemistry* 167:189-198.

Elchazly, M., and B. Thomas. 1976. Uber eine biochemische methode zur bestimmen der ballaststoffe und ihrer komponenten in pflanzlichen Lebensmitteln. *Zeitschrift fuer lebensmittel-Untersuchung und—Forschung* 162:329-340.

Englyst, H.N. 1981. Determination of carbohydrate and its composition in plant materials. In *The analysis of dietary fiber in food*. ed. W.P.T. James and O. Theander. pp. 71-93. New York: Marcel Dekker.

Englyst, H.N. 1985. Dietary polysaccharide breakdown in the gut of man. *Ph.D. thesis*. University of Cambridge, England.

Englyst, H.N. 1989. Classification and measurement of plant polysaccharides. *Animal Foods Science and Technology*. 23:27-42.

Englyst, H.N., V. Anderson, and J.H. Cummings. 1983. Starch and nonstarch polysaccharides in some cereal foods. *Journal of the Science of Food and Agriculture* 34:1434-1440.

Englyst, H.N., and J.H. Cummings. 1984. Simplified method for the measurement of total nonstarch polysaccharides in plant foods by gas-liquid chromatography of constituent sugars as alditol acetates. *Analyst* 109:937-942.

Englyst, H.N., and J.H. Cummings. 1985. Digestion of the polysaccharides of some cereal foods in the human small intestine. *American Journal of Clinical Nutrition* 42:778-787.

Englyst, H.N., and J.H. Cummings. 1986. Digestion of the carbohydrates of banana (Musa paradisiaca sapientum) in the human small intestine. *The American Journal of Clinical Nutrition* 44:42-50.

Englyst, H.N., and J.H. Cummings. 1988. An improved method for the measurement of dietary fibre as the nonstarch polysaccharides in plant foods. *Journal of the Association of Official Analytical Chemists* 71:808-814.

Englyst, H.N., and G.J. Hudson. 1987. Colorimetric method for routine measurement of dietary fibre as nonstarch polysaccharides. *Food Chemistry* 24:63-67.

Englyst, H.N., G.T. Macfarlane, and J.H. Cummings. 1988. New concepts in starch digestion in man. *Proceedings of the Nutrition Society* 47:64A.

Englyst, H.N., H. Trowell, D.A.T. Southgate, and J.H. Cummings. 1987. Dietary fiber and resistant starch. *American Journal of Clinical Nutrition* 46:873-874.

Englyst, H.N., H.S. Wiggins, and J.H. Cummings. 1982. Determination of the nonstarch polysaccharides in plant foods by gas-liquid chromatography of constituent sugars as alditol acetates. *Analyst* 107:307–318.

Faulks, R.M., and S.B. Timms. 1985. A rapid method for determining the carbohydrate component of dietary fibre. *Food Chemistry* 17:273–287.

Fibre 90—Chemical and Biological Aspects of Dietary Fibre. 1990. Royal Society of Chemistry—Food Chemistry Group. Norwich, United Kingdom.

Frølich, W., and M. Nyman. 1988. Minerals, phytate and dietary fibre in different fractions of oat-grain. *Journal of Cereal Science* 7:73–82.

Furda, I. 1981. Simultaneous analysis of soluble and insoluble dietary fiber. In *The Analysis of Fiber in Food*. ed. W.P.T. James and O. Theander, pp. 163–172. New York: Marcel Dekker.

Furda, I. 1989. Complexity of dietary fiber analysis. In *Frontiers of carbohydrate research: Food applications*. ed. R.P. Millane, J.N. BeMiller, and R. Chandrasekaran, pp. 83–98. London and New York: Elsevier Applied Science.

Furda, I., and C.S. Brine. 1990. eds. *New developments in dietary fiber*. New York: Plenum Press.

Galambos, J.T. 1967. The reaction of carbazole with carbohydrates. I. Effect of borate and sulfamate on the carbazole color of sugars. *Analytical Biochemistry* 19:119–132.

Goering, H.K., and P.J. Van Soest. 1970. *Forage Fiber Analysis*. Agriculture Handbook No. 397. Washington, D.C.: U.S. Department of Agriculture.

Graham, H., M.-B.G. Rydberg, and P. Åmen. 1988. Extraction of soluble dietary fiber. *Journal of Agricultural and Food Chemistry* 36:494–497.

Hall, J.M. 1989. A review of total dietary fiber methodology. *Cereal Foods World* 34:526–528.

Heaton, K.W. 1983. Dietary fibre in perspective. *Human Nutrition: Clinical Nutrition* 37c:151–170.

Hipsley, E.H. 1953. Dietary "fibre" and pregnancy toxaemia. *British Medical Journal* 2:420–422.

International Symposium. 1982a. *Dietary Fibre*. National College of Food Technology, University of Reading, Weybridge, Surrey, England.

International Symposium. 1982. *Fibre in Human and Animal Nutrition*. Massey University, Palmerston North, New Zealand.

Jeraci, J.L., B.A. Lewis, P.J. Van Soest, and J.B. Robertson. 1989. Urea enzymatic dialysis procedure for determination of total dietary fiber. *Journal of the Association of Official Analytical Chemists* 72:677–681.

Lee, S.C. 1989. Personal communication.

Lee, S.C., L. Prosky, and J.W. DeVries. 1991. Determination of soluble/insoluble and total dietary fiber in foods: collaborative study. *Journal of the Association of Official Analytical Chemists*. In press.

Li, B.W., and K.W. Andrews. 1988. Simplified method for determination of total dietary fiber in foods. *Journal of the Association of Official Analytical Chemists*. 71:1063–1064.

Lopez-Guisa, J.M., M.C. Harned, R. Dubielzig, S.C. Rao, and J.A. Marlett. 1988.

Processed oat hulls as potential dietary fiber sources in rats. *Journal of Nutrition* 118:953-962.

Madson, M., and M. Feather. 1978. Acidic decarboxylation of D-xyluronic, D-galacturonic and D-glycero-D-gulo-hepturonic acid. *176th American Chemical Society National Meeting # 61*. Miami Beach, FL.

Marabou food and fibre, Symposium. 1976. Marabou, Sundbyberg, Sweden. Reprinted in *Nutrition Reviews* 35:4-72, 1977.

Marlett, J.A. 1989. Measuring dietary fiber. *Animal Feed Science and Technology* 23:1-13.

Marlett, J.A. and D. Navis. 1988. Comparison of gravimetric and chemical analyses of total dietary fiber in human foods. *Journal of Agricultural and Food Chemistry* 36:311-315.

McQueen, R.E., and J.W.G. Nicholson. 1979. Modification of the neutral detergent fiber procedure for cereals and vegetables by using α-amylase. *Journal of the Association of Official Analytical Chemists* 62:676-680.

Meuser, F., P. Suckow, and W. Kulikowski. 1983. Analytische bestimmung von ballastoffen in broten. Obst und gemuse. *Getreide Mehl und Brot* 37:380-383.

Meuser, F., P. Suckow, and M. Kulikowski. 1985. Verfahren zur bestimmung von unloslichen und loslichen ballastoffen in lebenmitteln. *Zeitschrift fuer Lebensmittel- Untersuchung und - Forschung* 181:101-106.

Mongeau, R., and R. Brassard. 1979. Determination of neutral detergent fiber, hemicellulose, cellulose and lignin in breads. *Cereal Chemistry* 56:437-441.

Mongeau, R., and R. Brassard. 1986. A rapid method for the determination of soluble and insoluble dietary fiber: comparison with AOAC total dietary fiber procedure and Englyst's method. *Journal of Food Science* 51:1333-1336.

Mongeau, R., and R. Brassard. 1990b. A comparison of three methods for analyzing dietary fiber in 38 foods. *Journal of Food Composition and Analysis* 2:189-199.

Mongeau, R., and R. Brassard. 1990a. Determination of insoluble, soluble, and total dietary fiber: Collaborative study of a rapid gravimetric method. *Cereal Foods World* 35:319-322.

Morrison, I.M. 1980. Hemicellulosic contamination of acid detergent residues and their replacement by cellulosic residues in cell wall analysis. *Journal of the Science of Food and Agriculture* 23:455-463.

Mugford, D.C. 1991. Fibre methods for cereal foods: an Australasian interlaboratory survey. *Food Australia*. In press.

Nyman, M., K.-E. Palsson, and N.-G. Asp. 1987. Effects of processing on dietary fiber in vegetables. *Lebensmittel-Wissenschaft und Technologie* 20:29-36.

Painter, N.S. 1975. *Diverticular Disease of the Colon: A Deficiency Disease of Western Civilization*. London: Heinemann.

Paul, A.A., and D.A. Southgate. 1978. In *McCance and Widdowson's The Composition of Foods*. 4th Edition, pp. 8-9, London: HM Stationery Office.

Prosky, L. 1981. Discussion and the definition and analysis of fibre. *Association of Official Analytical Chemists Spring Workshop*. Ottawa, Canada.

Prosky, L., N.-G. Asp, I. Furda, J.W. DeVries, T.F. Schweizer, and B.F. Harland. 1984. Determination of total dietary fiber in foods and food products and total diets: interlaboratory study. *Journal of the Association of Official Analytical Chemists* 67:1044-1053.

Prosky, L., N.-G. Asp, I. Furda, J.W. Devries, T.F. Schweizer, and B.F. Harland. 1985. Determination of total dietary fiber in foods and food products: collaborative study. *Journal of the Association of Official Analytical Chemists* 68:677-679.

Prosky, L., N.-G. Asp, T.F. Schweizer, J.W. DeVries, and I. Furda. 1988. Determination of insoluble, soluble, and total dietary fiber in foods and food products: interlaboratory study. *Journal of the Association of Official Analytical Chemists* 71:1017-1023.

Prosky, L., N.-G. Asp, T.F. Schweizer, J.W. DeVries, and I. Furda. 1991. Determination of insoluble and soluble dietary fiber in foods and food products: collaborative study. *Journal of the Association of Official Analytical Chemists.* In press.

Prosky, L., and B.F. Harland. 1979. Need, definition, and rationale regarding dietary fiber. *Association of Official Analytical Chemists 93rd Annual Meeting.*

Prosky, L., and B.F. Harland. 1981. Definition and method for dietary fiber. *Association of Official Analytical Chemists 95th Annual Meeting, Abstract 63.*

Prosky, L., and B.F. Harland. 1982. Collaborative study of a method for dietary fiber. *Association of Official Analytical Chemists 96th Annual Meeting, Abstract 103.*

Rabe, E. 1987. Bestimmung der unloslichen und loslichen ballastroffe. *Getreide, Mehl und Brot* 41:297-305.

Rabe, E., W. Seibel, P. Sukow, and F. Meuser. 1988. Vergleichende bestimmung von unloslichen, loslichen und gesamtballastroffen in getreide erzeugnissen. *Getreide, Mehl und Brot* 42:297-305.

Rasper, V.F., J.M. Brillouet, D. Bertrand, and C. Mercier. 1981. Analysis of dietary fiber in faeces of rats fed with fiber-supplemented diets. *Journal of Food Science* 46:559-563.

Robertson, J.B., and P.J. Van Soest. 1977. Dietary fiber estimation in concentrate feed-stuffs. *Journal of Animal Science* 45(supplement):254-255.

Robertson, J.B., and P.J. Van Soest. 1981. The detergent system of analysis and its application to human foods. In *The Analysis of Dietary Fiber in Foods*. ed. W.P.T. James and O. Theander, pp. 123-158. New York: Marcel Dekker.

Roe, J.H. 1955. The determination of sugar in the blood and spinal fluid with anthrone reagent. *Journal of Biological Chemistry* 212:335-343.

Roth, H.P., and M.A. Mehlman. 1978. Role of dietary fiber in health. *American Journal of Clinical Nutrition Supplement* 31:S1-S291.

Saeman, J.F., W.E. Moore, and M.A. Millet. 1963. Sugar units present. Hydrolysis and quantitative paper chromatography. In *Methods in carbohydrate chemistry* vol. 3, chapter 12, ed. R.L. Whistler, pp. 54-69. New York: Academic Press.

Schweizer, T.F., H. Andersson, A.M. Langkilde, S. Reimann, and I. Torsdottir. 1990. Nutrient excreted in ileostomy effluents after consumption of mixed diets with beans or potatoes. II. Starch, dietary fiber and sugars. *European Journal of Clinical Nutrition* 44:567-575.

Schweizer, T.F., W. Frøliche, S. Del Vedovo, and R. Besson. 1984. Minerals and phytate in the analysis of dietary fiber from cereals. *I. Cereal Chemistry* 61:116-119.

Schweizer, T.F., E. Walter, and P. Venetz. 1988. Collaborative study for the enzymatic, gravimetric determination of total dietary fiber in foods. *Mitteilungen aus dem Gebiete der Lebensmitteluntersuchung und Hygiene* 79:57-68.

Schweizer, T.F., and P. Würsch. 1979. Analysis of dietary fibre. *Journal of the Science of Food and Agriculture* 30:613-619.

Schweizer, T.F., and P. Würsch. 1981. Analysis of dietary fiber. In *The Analysis of Dietary Fiber in Foods*, ed. W.P.T. James and O. Theander, pp. 203-216. New York: Marcel Dekker.

Selvendran, R.R., and M.S. DuPont. 1984. Problems associated with the analysis of dietary fibre and some recent developments. *Development in Food Analysis Techniques* 3:1-68.

Selvendran, R.R., J.F. March, and S.G. Ring. 1979. Determination of aldoses and uronic acid content of vegetable fibre. *Analytical Biochemistry* 96:282-292.

Selvendran, R.R., A.V.F.W. Verne, and R.M. Faulks. 1989. Methods for analysis of dietary fibre. In *Modern Methods of Plant Analysis*, ed. H.F. Linskens and J.F. Jackson, pp. 234-259. Berlin, Heidelberg, New York, London, Paris, Tokyo: Springer-Verlag.

Shinnick, F.L., M.J. Longacre, S.L. Ink, and J.A. Marlett. 1988. Oat fiber: composition versus physiological function in rats. *Journal of Nutrition* 118:144-151.

Siljeström, M., and N.-G. Asp. 1985. Resistant starch formation during baking—effect of baking time and temperature and variations in the recipe. *Zeitschrift fuer Lebensmittel Untersuchung und Forschung* 181:4-8.

Southgate, D.A.T. 1969. Determination of carbohydrates in foods. II. Unavailable carbohydrates. *Journal of the Science of Food and Agriculture* 20:331-335.

Southgate, D.A.T. 1981. Use of the Southgate method for unavailable carbohydrates in the measurement of dietary fiber. In *The Analysis of Dietary Fiber in Foods*. ed. W.P.T. James and O. Theander, pp. 1-19. New York: Marcel Dekker.

Southgate, D.A.T. 1982. Definition and terminology of dietary fiber. In *Dietary Fiber in Health and Disease*, ed. G.V. Vahouny and D. Kritchevsky, pp. 1-7. New York: Plenum Press.

Spiller, G.A., and R.M. Kay. 1979. Recommendation and conclusions of the dietary fiber workshop of the XI International Congress of Nutrition, Rio de Janeiro, 1978. *American Journal of Clinical Nutrition* 32:2102-2103.

Theander, O. 1983. Advances in the chemical characterization and analytical determination of dietary fibre components. In *Dietary Fibre*, ed. G.G. Birch, and K.J. Parker, pp.77-93. London: Applied Science Publishers.

Theander, O. and P. Åmen. 1979. Studies on dietary fibres. 1. Analysis and chemical characterization of water-insoluble dietary fibres. *Swedish Journal of Agricultural Research* 9:99-106.

Theander, O. and P. James. 1979. European efforts in dietary fiber characterization. In *Dietary Fibres: Chemistry and Nutrition*, G.E. Inglett and I. Falkehag, pp. 245–249. New York and London: Academic Press.

Theander, O., and E. Westerlund. 1986a. Studies on dietary fiber. 3. Improved procedures for analysis of dietary fiber. *Journal of Agricultural and Food Chemistry* 34:330–336.

Trowell, H.C. 1972a. Dietary fibre and coronary heart disease. *Revue Europeene d'Etudes Cliniques et Biologique* 17:345–349.

Trowell, H.C. 1972b. Ischemic heart disease and dietary fiber. *American Journal of Clinical Nutrition* 25:926–932.

Trowell, H. 1972c. Crude fibre, dietary fibre and atherosclerosis. *Atherosclerosis* 16:138–140.

Trowell, H.C. 1974. Definitions of fibre. *Lancet* 1:503.

Trowell, H.C. 1976. Definition of dietary fiber and hypotheses that it is a protective factor in certain diseases. *American Journal of Clinical Nutrition* 29:417–427.

Trowell, H., D.A.T. Southgate, T.M.S. Wolever, A.R. Leeds, M.A. Gassull, and D.J.A. Jenkins. 1976. Dietary fiber redefined. *Lancet* 1:967.

U.S. Department of Agriculture and U.S. Department of Health and Human Services. 1980. *Nutrition and your health. Dietary guidelines for Americans.* USDA, DHHS, Washington, D.C.

Van Soest, P.J. 1963. Use of detergents in the analysis of fibrous feeds. I. Preparation of fiber residues of low nitrogen content. *Journal of the Association of Official Agricultural Chemists* 46:825–829.

Van Soest, P.J. 1963. Use of detergents in the analysis of fibrous feeds: II. A rapid method for the determination of fiber and lignin. *Journal of the Association of Official Analytical Chemists* 46:829–835.

Van Soest, P.J. 1973. Collaborative study of acid-detergent fiber and lignin. *Journal of the Association of Official Analytical Chemists* 56:781–784.

Van Soest, P.J., and R.W. McQueen. 1973. The chemistry and estimation of fibre. *The Proceedings of the Nutrition Society* 32:123–130.

Van Soest, PJ., and R.H. Wine. 1967. Use of detergents in the analysis of fibrous feeds. IV. Determination of plant cell-wall constituents. *Journal of the Association of Official Analytical Chemists* 50:50–55.

Varo, P., R. Laine, and P. Koivistoinen. 1983. Effect of heat treatment on dietary fibre: interlaboratory study. *Journal of the Association of Official Analytical Chemists* 66:933–938.

Walker, A.R. 1947. The effects of recent changes of food habits on bowel motility. *South African Medical Journal* 21:590–596.

Westerlund, E. In Press.

Westerlund, E., O. Theander, R. Andersson, and P. Åman. 1989. Effect of baking on polysaccharides in white bread fractions. *Journal of Cereal Science* 10:149–156.

3

Regulations and Marketing

In 1987, the U.S. Food and Drug Administration (FDA) amended its food labeling regulations, to exclude nondigestible dietary fiber when determining the calorie content of food for nutrition labeling purposes, to allow for more accurate declaration of the available calories in food. In that Final Rule (1987), it was stated that the nondigestible dietary fiber may be subtracted from the total carbohydrate content before calculating the calories contributed by the carbohydrate fraction of the food. It further stated that the method to be used for determining the indigestible dietary fiber would be "Total Dietary Fiber in Food: Enzymatic—Gravimetric Method. First Action in the Journal of the Association of Official Analytical Chemists (1985) as amended in that same journal in 1986." The method chosen by the FDA was one that was relatively simple, could be performed by an inexperienced chemist, did not require large expenditures for equipment or reagents, and had successfully passed an AOAC collaborative study.

The method is an enzymatic-gravimetric procedure, outlined in Figure 3-1 below (Prosky et al. 1984).

At the Spring 1981 AOAC Workshop in Ottawa, the consensus reached, after considerable discussion, was that dietary fiber for nutritional labeling purposes should be that fraction of the food sample that is not digestible under the conditions of the test, which was to be collaboratively studied. The test would replicate the enzymatic digestion of the human G.I. tract as closely as possible, yet be practical enough to be carried out in any competent laboratory. To ensure that the enzymes being used were in fact adequate for digesting naturally occurring starches and proteins under the conditions of the test method, a procedure was estab-

Sample preparation, dried pulverized, and extracted with 25 ml petroleum ether to remove lipids.

↓

Extraction repeated two times with 25 ml petroleum ether. Weighing of duplicate samples, 1 g. Addition of 50 ml phosphate buffer, pH 6.0. Addition of Termamyl for gelatinization, 100 μl.

↓

Incubation, boiling water bath, 15 min, shaking at 5-min intervals. Adjustment to pH 7.5 with 0.285 N NaOH.

↓

Addition of protease, 5 mg. Incubation, 60°C, 30 min, continuous agitation. Adjustment to pH 4.5 with 10 ml 0.329 M phosphoric acid. Addition of amyloglucosidase, 0.3 ml. Incubation, 60°C, 30 min, continuous agitation. Addition of 280 ml 95% ethyl alcohol, preheated to 60°C.

↓

Formation of precipitate, 60 min. Filtration through bed containing 0.5 g Celite 545. Washed with three 20-ml portions of 74% ethyl alcohol, two 10-ml portions of 95% ethyl alcohol, and two 10-ml portions of acetone.

↓

Drying of crucibles, 70°C vacuum oven or 105°C air oven, overnight. Cooling and weighing of crucibles.

Protein determination, use entire crucible contents. Factor = 6.25 × nitrogen.

Ash determination 525°C, 5 h, cool and weigh.

FIGURE 3-1. Calculation of % TDF = weight of residue, minus weight of protein, minus weight of ash, minus weight of blank.

lished for validating the efficacy of the enzyme treatment. This validation consisted of analyzing various starches and proteins with the method and confirming the absence of significant recoverable residue. Further assurance against undesirable activity was achieved by testing on various substrates and assuring complete recoveries. This approach is still proper.

REGULATIONS

First Study AOAC

The first study of this method was an interlaboratory study and was conducted to determine the total dietary fiber (TDF) content of food, food products, and total diets, using the method. Thirteen unknown products, including two mixed diets (one lacto-ovo vegetarian and the other non-vegetarian), were analyzed by 32 analysts from 14 different countries. Duplicate samples of dried foods were gelatinized with Termamyl, a heat-stable alpha-amylase, and then enzymatically digested with protease and amyloglucosidase, to remove the protein and starch present in the sample. Four volumes of 95% ethanol were then added to precipitate the soluble dietary fiber. The total residue was filtered and then washed with 74% ethanol, 95% ethanol, and acetone. After drying, the residue was weighed. One of the sample duplicates was analyzed for protein, and the other was ashed at 525° C and the ash was measured. TDF was calculated as the weight of the residue less the weight of protein and ash, corrected by a blank, of course. Coefficients of variation (CV) for ten of the samples ranged from 2.95% to 26.39%, depending on the concentration. For three of the products, high coefficients of variation were obtained. The results compared satisfactorily with those obtained previously by the best method available for the individual foods studied.

Second Study AOAC

The second study, again an interlaboratory collaborative, was conducted by basically the same method, with minor changes in the concentration of alcohol and buffers, incubation time, sample preparation, and some explanatory notes, all with the intent of decreasing the coefficient of variation of the method. Duplicate blind samples of soy isolate, white wheat flour, potatoes, rice, wheat bran, rye bread, oats, corn bran, and whole wheat flour were analyzed by nine collaborators in five countries.

Discussion of this method at the 2nd International Symposium on Dietary Fiber, held in Washington, D.C., showed that it was one of the two most widely accepted procedures for determining TDF (Southgate 1986). In this collaborative study, we sought to: (1) determine at what fiber levels the method had an unacceptably high level of reproducibility; (2) assess the extent of enzymatic starch digestion because incomplete removal of the starch would interfere with the determination of TDF; (3) evaluate the optimum mesh size for grinding samples before determining TDF; and (4) improve the CV values for determining TDF by clarifying directions and altering some of the concentrations of the solutions. The results of the individual analysis are shown in Table 3-1.

As can be seen, excellent results were obtained with all samples except soy isolate and rice, matrices very low in dietary fiber and high in protein and starch. For the soy isolate, two laboratory values were discarded

TABLE 3-1. **Collaborative results (blind duplicates) of determination of TDF by the enzymatic-gravimetric method. TDF is expressed as % dry weight.**

Coll.	Soy Isolate	White Wheat Flour	Rye Bread	Potatoes	Rice	Wheat Bran	Oats	Corn Bran	Whole Wheat Flour
1	3.18[a]	2.40	6.22	5.41[b]	0.91	42.32	10.30	85.14	10.54
	0.00	2.41	6.51	5.75	0.35	41.86	11.66	86.32	11.96
2	0.73	2.63	5.61[a]	7.30	0.42[a]	40.64	9.99	83.81[c]	12.02
	—[d]	3.08	8.07	8.08	2.18	41.41	10.93	84.04	11.49
3	1.34	2.40	6.75	6.02	0.54	42.79	11.30	86.90	13.48
	1.23	2.40	7.05	6.70	0.55	42.32	11.79	87.96	13.15
4	1.19	2.79	6.30	7.85	1.28	44.01	10.98	86.70	12.70
	1.18	2.78	6.40	6.94	1.14	41.24	10.65	86.76	12.28
5	1.17	2.90	7.02	7.19	0.74	44.38	10.87	87.21	12.62
	0.00	2.91	6.35	7.38	0.82	41.75	11.25	86.80	12.43
6	0.75	2.67	6.11	7.06	0.90	42.34	10.98	87.59	13.03
	0.78	3.02	6.44	6.81	1.42	42.19	10.29	87.84	13.62
7	2.52	3.09[b,e]	6.60	6.99	0.27	43.43	12.99[b]	88.25	12.58
	—[d]	8.42	6.34	7.15	0.77	44.31	12.48	87.89	13.15
8	2.25	3.14	6.88	7.22	1.91	43.60	11.80	87.31	13.50
	1.99	3.00	7.34	7.90	1.74	44.20	10.75	88.42	12.68
9	2.70	3.08	6.35	7.97	1.49	41.96	11.70	87.47	13.11
	1.76	2.89	6.58	7.51	1.26	43.00	11.05	87.14	12.91

[a]Cochran outlier, results used
[b]Cochran outlier, results eliminated
[c]Grubbs outlier, results used
[d]Erroneous results due to filtration problem
[e]Grubbs outlier, results eliminated

Reprinted from Prosky et al. *Journal of the Association of Official Analytical Chemists* 1985. 68: 678.

because of filtration problems and one was a Cochran outlier (Prosky et al. 1985). The Cochran outlier was used in the statistical evaluation because including the value did not significantly alter the CV. For white wheat flour, the results of Collaborator 7 were not used in the final statistical analysis of the data because the value was a Cochran and Grubbs outlier. The TDF values for potatoes (Collaborator 1) and for oats (Collaborator 7) were not used in the statistical analyses; both were Cochran outliers. All TDF values were used in the final statistics for rye bread, rice, wheat bran, corn bran, and whole wheat flour.

The measures of precision for determining TDF are shown in Table 3-2. The seven food samples having TDF values of >2.78% had CV values of <10%. When the foods contained <1.5% fiber, the CV was about 66%. This procedure had substantially reduced the CV values of the food samples, when compared with the previous interlaboratory study (Prosky et al. 1984). The two samples with low-fiber content, namely, 1.42% and 1.04% for soy isolate and rice, which have large CV values, may indeed be unimportant because of their low-fiber content. On the other hand, the method gave good results at the 2.78% dietary fiber level (white wheat flour) and higher.

The changes in the time of enzyme incubation, concentration of buffers and reagents, and directions for carrying out the procedure have made the TDF method more rugged.

Among the samples analyzed in the second interlaboratory study (Prosky et al. 1985), both potatoes and white flour had very high starch contents, yet the enzyme digested all of the starch. Lack of homogeneity of the rice sample in the previous collaborative study had made the starch in that sample unavailable for hydrolysis. In the 1985 collabora-

TABLE 3-2. Measures of precision for determining TDF.

Sample	Dietary Fiber % dry basis	Repeatability CV, %	Reproducibility CV, %
Soy isolate	1.42	66.25	66.25
White wheat flour	2.78	5.55	9.80
Rye bread	6.58	3.94	5.29
Potatoes	7.25	5.66	7.49
Rice	1.04	45.62	53.71
Wheat bran	42.65	2.33	2.66
Oats	11.03	5.30	5.30
Corn bran	86.86	0.56	1.56
Whole wheat flour	12.57	3.67	5.92

Reprinted from Prosky et al., *The Journal of the Association of Official Analytical Chemists* 68:679 (1985).

tive study, the mesh size was better defined; therefore, the matrix played no role in the enzymatic procedure. The enzymatic-gravimetric method for determining TDF was adopted official first action in 1986 and official final action in 1987 by the AOAC.

Third Study AOAC

A third collaborative study (Prosky et al. 1988) was conducted on a modified version of the adopted official method, to determine the insoluble dietary fiber (IDF), soluble dietary fiber (SDF), and total dietary fiber (TDF) content of food and food products. Considering that IDF and SDF often exhibit distinct physiological effects (Spiller 1986), the basic method was extended to give separate values for SDF and IDF, if desired, as well as TDF (Asp et al. 1983).

The method employed was basically the same as the one developed for TDF only, which was adopted official final action by the AOAC, but which contained an added step for separating the insoluble from the soluble dietary fiber. In addition, changes were made in the concentration of buffer and base, and hydrochloric acid was substituted for phosphoric acid. This collaborative study used the same three enzymes; the concentration of the phosphate buffer was increased from 0.05M to 0.08M, the concentration of sodium hydroxide was increased from 0.17N to 0.275N, and a 0.325N hydrochloric acid solution was substituted for the 0.205M phosphoric acid solution. A collaborative study conducted in Switzerland during this same time frame (Schweizer et al. 1988) gave excellent results for TDF, using the AOAC methods, with the above changes included.

The changes were introduced into the adopted method to improve its robustness. The increased buffering capacity of the initial incubation mixture avoids pH adjustments when acidic products (for example, fruit pulps) are analyzed. The change in the acid used became necessary in order to avoid a concomitant increase in the final phosphate concentration, which had given rise to salt coprecipitation (Prosky et al. 1984). The authors also hoped to establish methods for SDF and IDF, in addition to those established for TDF, by filtering the IDF before precipitating the SDF with ethanol. We sought to determine the agreement between the independent method for TDF, compared with TDF derived by summing SDF and IDF. Furthermore, the calculations were simplified by using new data sheets and equations. Figure 3-2 shows the new blank and test sample data sheets and equations for calculating total dietary fiber. If one wanted to calculate SDF and IDF, one would use for the mean blank $a+c/2$ and $b+d/2$, respectively.

AOAC DIETARY FIBER METHOD BLANK DATA SHEET

BLANKS	a		b		c		d	
Crucible + Celite + tare Weight (mg)								
Crucible + Celite + Residue Weight (mg)								
Residue Weight (mg)	R_1	R_2	R_1	R_2	R_1	R_2	R_1	R_2
Protein (mg) P								
Crucible + Celite + Ash weight (mg)								
Ash weight (mg) A								
Blanks (mg)								
Mean Blank (a+b+c+d)/4 (mg) B								

$$\text{Blanks (mg)} = \frac{R_1 + R_2}{2} - P - A$$

AOAC DIETARY FIBER METHOD SAMPLE DATA SHEET

	Sample				Sample			
	1		2		1		2	
1. Sample Weight (mg)	m_1	m_2	m_1	m_2	m_1	m_2	m_1	m_2
2. Crucible + Celite tare weight (mg)								
3. Crucible + Celite + Residue weight (mg)								
4. Residue Weight (mg)	R_1	R_2	R_1	R_2	R_1	R_2	R_1	R_2
5. Protein (mg) P								
6. Crucible + Celite + Ash weight (mg)								
7. Ash Weight (mg) A								
8. Mean Blank (mg) B								
9. Dietary Fiber (%)								

$$\text{Dietary Fiber (\%)} = \frac{\frac{R_1 + R_2}{2} - P - A - B}{\frac{m_1 + m_2}{2}} \times 100$$

FIGURE 3-2. Sample and blank data sheet and equations for calculating blank and percent fiber. Reprinted with permission of Prosky et al. 1988. The *Journal of the Association of Official Analytical Chemists* 71: 1019.

Duplicate blind samples were analyzed by 13 collaborators. These collaborators were analysts in food companies, universities, and commercial and government laboratories, representing seven countries. Collaborators were sent ten duplicate blind samples for analysis: (1) soy protein isolate, Supro 610K, Lot No. C3A BK-006, donated by Ralston Purina Co., St. Louis, MO; (2) white wheat flour, low extraction (0.45% ash, 12% protein), donated by General Mills, Inc., Minneapolis, MN; (3) rye bread, Deli Rye (Giant Foods, Inc., Washington, D.C.), dried for 4 hours at 80°C; (4) potatoes, instant (Giant Foods); (5) rice, enriched long-grain (Giant Foods); (6) corn bran, Lot No. CFL 2601A, donated by A.E. Staley Manufacturing Co., Decatur, IL; (7) oats, quick cooking, donated by Quaker Oats Co., Barrington, IL; (8) Fabulous Fiber, a mixture of malto dextrin, whey, psyllium hulls, guar gum, pectins, vitamins, and minerals (Lewis Laboratories International, Ltd, Westport, CT); (9) wheat bran, AACC certified food grade, purchased from AACC, St. Paul, MN; (10) high-fiber cereal AL T., donated by Farma Food, Inc., Washington, D.C. The rye bread, potatoes, rice, oats, wheat bran, and high-fiber cereal were ground to a uniform size of 350 um in a Microjet 10 centrifugal mill (Quartz Technology, Inc., Westbury, NY). No heating of the products occurred during this procedure.

Test samples were placed in plastic bags with an identifying letter taped to each bag. None of the test samples had >10% fat; therefore, fat extraction was not recommended. Each test sample was to be dried by the participating laboratory at 70°C in a vacuum oven (preferred method), or overnight in a 105°C air oven, and stored in a desiccator until analyzed.

The measures of IDF, SDF, and TDF measured independently are shown in Tables 3-3, 3-4, and 3-5. In these tables, RSD_r is the repeatability relative standard deviation and is a measure (in percent) of how well a typical laboratory can consistently obtain the same results under the same conditions. In other words, RSD_r measures within-laboratory variability. RSD_R is the reproducibility relative standard deviation and is a measure of among-laboratory variability. The Fabulous Fiber sample had a reproducibility coefficient of 49% for IDF (Table 3-3); however, this was based on data from only four laboratories. The paucity of results was mainly due to the filtration problems associated with this partly soluble, viscous, but thixotropic sample. The high coefficients of reproducibility RSD_R for rice and soy isolate were due mainly to the low IDF in these products. The fact that the IDF for soy isolate was higher than the TDF measured independently suggests that, when the sample was prepared for IDF determination, another substance in addition to insoluble fiber may not have been going into aqueous solution, but was soluble in 80% ethanol. The other

TABLE 3-3. Measures of precision for determining IDF.

Product	Average IDF, % dry wt	No. of Collaborators	Repeatability RSD_r, %	Reproducibility RSD_R, %
Corn bran	87.47	11	0.38	0.87
Fabulous Fiber	9.13	4	7.31	48.79
High fiber cereal	30.14	11	4.24	4.33
Oats	5.66	9	11.11	12.07
Potatoes	4.85	11	7.25	11.94
Rice	0.75	11	28.38	41.52
Rye bread	5.42	11	6.66	15.67
Soy isolate	6.31	10	15.29	28.56
Wheat bran	41.59	11	1.98	3.40
White wheat flour	1.56	10	11.57	21.03

Reprinted from Prosky et al., the *Journal of the Association of Official Analytical Chemists* 71:1022 (1988).

foods gave reasonable results, with the reproducibility inversely proportional to the IDF content of the product.

Table 3-4 shows the measures of precision for determining SDF. The reproducibility coefficients were very high for rice and soy isolate (128 and 100%, respectively) and in the 20 to 50% range for the remaining foods except corn bran (RSD_R 79%). The high variability of reproducibility is mainly due to the very low SDF content of the foods, the exception being Fabulous Fiber, which had about 9% SDF and a 45% coefficient of reproducibility. The problem with filtering this product has been discussed earlier. It is nonetheless noteworthy that the variability in results in

TABLE 3-4. Measures of precision for determining SDF.

Product	Average SDF, % dry wt	No. of Collaborators	Repeatability RSD_r, %	Reproducibility RSD_R, %
Corn bran	0.04	11	40.22	78.57
Fabulous Fiber	8.95	6	8.07	44.72
High fiber cereal	1.88	11	32.64	34.74
Oats	4.21	11	10.78	26.73
Potatoes	2.14	11	19.27	27.88
Rice	0.19	11	100.03	127.52
Rye bread	1.54	11	13.19	28.67
Soy isolate	0.46	10	71.86	100.85
Wheat bran	2.87	11	12.01	19.78
White wheat flour	1.17	9	23.52	23.52

Reprinted from Prosky et al., the *Journal of the Association of Official Analytical Chemists* 71:1022 (1988).

TABLE 3-5. Measures of Precision for Determining TDF Independently.

Product	Average TDF, % dry wt	No. of Collaborators	Repeatability RSD_r, %	Reproducibility RSD_R, %
Corn bran	87.13	9	0.55	1.62
Fabulous Fiber	16.83	7	3.79	11.19
High fiber cereal	32.18	9	2.63	3.98
Oats	11.31	8	6.51	14.32
Potatoes	6.78	9	9.15	13.93
Rice	1.01	9	54.32	70.42
Rye bread	6.61	9	11.14	12.20
Soy isolate	1.57	9	41.81	88.02
Wheat bran	43.87	9	2.14	2.70
White wheat flour	3.09	7	13.58	13.58

Reprinted from Prosky et al., the *Journal of the Association of Official Analytical Chemists* 71:1022 (1988).

either IDF or SDF compensated for each other, and the sums were confined to a narrower range. Thus, for Fabulous Fiber, the RSD_R of the sum (that is, total dietary fiber by adding IDF to SDF) would be only 8%, compared with over 40% for the individual fractions.

Table 3-5 shows the TDF independent analysis values and the measures of precision. All foods analyzed had 15% coefficient of reproducibility or better, with the exception of rice and soy isolate, which had RSD_R values of 70% and 88%, respectively. These two foods have always had large RSD_R values because of their low-fiber content (Prosky et al. 1984; 1985). While RSD_r and RSD_R for TDF in this study were not quite as good as those obtained in the earlier collaborative study, they were still acceptable. The increase in variability may have been due to the much heavier workload involved in this study (that is, three times as many crucibles had to be handled when compared to the earlier study). For the eight products previously collaboratively studied, the measures of precision were calculated with the elimination of only 2 of 142 results as outliers, whereas in the previous study, 6 of 160 results were eliminated as outliers. The average TDF values were nearly identical to those from an earlier study (Prosky et al. 1984). Thus, the changes in the buffer concentrations and the acid did not alter the performance of the method.

When the TDF by independent analysis was compared with the TDF obtained by summing TDF and SDF, the results were nearly identical (Figure 3-3). With the exception of soy isolate, which had four times as much IDF as TDF (due to the lack of solution of some material other than fiber, in the fractionation process) and therefore about four times as much TDF calculated by summing IDF and SDF, as compared with the independent determination of TDF, all other values for TDF obtained by

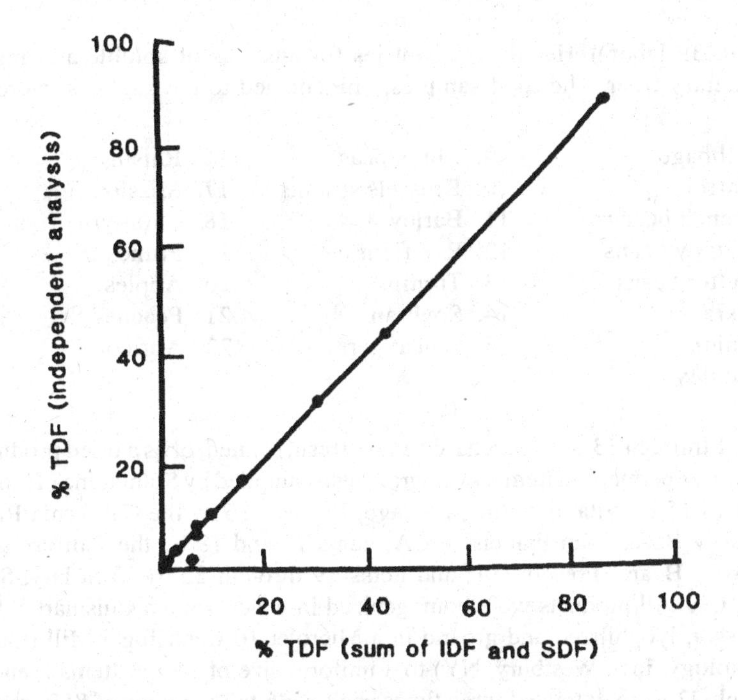

FIGURE 3-3. Percent TDF by independent analysis, plotted against the sum of insoluble and soluble dietary fiber for the ten products listed in Table 3-5.

summing IDF and SDF were acceptable. This eliminates the necessity of determining TDF independently, except when TDF alone is desired. However, the precision of the method for the individual fractions, especially for SDF, is not yet satisfactory. This is due in part to choosing products with very low SDF or with special problems.

It was recommended by the AOAC foods committee that the method for SDF and IDF required further study. The modifications (changing the concentration of buffer and base and using hydrochloric acid instead of phosphoric acid) to the official final action method for TDF were adopted.

Fourth Study AOAC

To carry out some of the recommendations of the AOAC foods committee, a fourth collaborative study was conducted on a larger number of products, representing a greater variety of matrices. Portions of 15 food samples (seven blind duplicates and one standard containing 4.3% to 5.4% insoluble dietary fiber and 1.5% to 2.7% soluble dietary fiber) were sent to 39 laboratories in 13 countries for analysis of soluble and insoluble dietary fiber. The food samples, unidentified to the analysts, were:

1. Cabbage
2. Carrots
3. French beans
4. Kidney beans
5. Butter beans
6. Okra
7. Onions
8. Parsley
9. Chick peas
10. Brussels sprouts
11. Barley
12. Rye flour
13. Turnips
14. Soybran
15. Wheat germ
16. Raisins
17. Mission figs
18. Calimyrna figs
19. Prunes
20. Apples
21. Peaches
22. Apricots

Items 1 through 13 were purchased either fresh, canned, or as a dried product at the local supermarket. Item 14 was graciously supplied by Solnut, Inc., Hudson, IA; item 15 by Vitamins, Inc., Chicago, IL; item 16 by the California Raisin Advisory Board, San Francisco, CA; items 17 and 18 by the California Fig Advisory Board, Fresno, CA; and items 19 through 22 by Vacu-Dry, Santa Rosa, CA. All products were homogenized in water using a Cuisinart™ food processor, lyophilized and ground in a Microjet 10 Centrifugal Mill (Quartz Technology, Inc., Westbury, NY) to a uniform size of 350 μ. Items 2 and 17 through 22 were extracted three times each, with ten volumes of 85% methanol, to remove sugars, so that the final material could be lyophilized to a dry material (otherwise they were sticky from the high sugar content). The final material was ground again in a Microjet 10 Centrifugal Mill, as described above. All test samples were placed in 25 ml scintillation vials, with screw caps, and an identification number was taped to each vial. The summary of results of the present study are shown in Tables 3-6 and 3-7. With most of the collaborators reporting their results, the measures of precision for determining IDF and SDF are shown in these two tables. Only two individual laboratory values were dropped because of statistical consideration. One of 15 laboratories was dropped in the determination of IDF in okra and 1 of 14, in the determination of IDF in kidney beans. All other values were used as reported by the analysts.

The average IDF (dry basis) ranged from 4.30% for barley to 65.24% for soybran (Table 3-6). The repeatability (RSD_r) of the determination for the 22 foods analyzed ranged from 0.86% for apricots to 10.38% for chick peas, which is considered excellent for this analysis. The variation among the laboratories (RSD_R) ranged from 3.68% for soybran to 19.44% for prunes, which was also very good when one considers that half the foods had RSD_R values of less than 10% and an additional seven foods had RSD_R values of less than 15%. The highest RSD_R's were found for prunes, raisins, and Calimyrna figs and the lowest for apples. The measures of precision for the

TABLE 3-6. Measures of Precision for Determining IDF.

Product	Average IDF, % dry wt	No. of Collaborators	Repeatability RSD_r, %	Reproducibility RSD_R, %
Cabbage	21.60	9	4.00	7.79
Carrots	32.29	12	5.38	11.39
French beans	25.64	10	3.23	5.87
Kidney beans	16.33	14	4.53	6.39
Butter beans	17.36	10	2.34	11.31
Okra	24.15	15	6.43	13.57
Onions	13.32	12	6.51	11.79
Parsley	34.39	12	3.56	13.64
Chick peas	16.69	12	10.38	16.80
Brussels sprouts	30.23	15	2.27	7.89
Barley	4.30	12	9.92	14.30
Rye flour	11.81	15	4.87	8.62
Turnips	21.83	12	6.60	16.61
Soybran	65.24	13	1.40	3.68
Wheat germ	15.67	9	4.54	6.13
Seedless raisins	49.18	8	5.51	19.30
Mission figs	33.61	6	2.76	12.09
Calimyrna figs	43.07	5	5.59	18.40
Prunes	46.18	6	6.11	19.44
Apples	55.57	4	0.92	4.55
Peaches	39.53	6	2.17	6.16
Apricots	44.92	5	0.86	8.22

determination of SDF are presented in Table 3-7. The products analyzed had an average SDF that ranged from 1.34% for chick peas to 33.42% for prunes. The repeatability (RSD_r) values varied from 2.34% for apples to 39.06% for chick peas, with approximately 50% of the reporting laboratories having less than a 10% variation in their results. The reproducibility (RSD_R) values showed that approximately 50% of the reporting laboratories had an RSD_R of more than 20%, 45% of the foods had RSD_R values of between 10% and 20%, and only one food (French beans) had an RSD_R value less than 10%. These preliminary results indicate that a major reason for the differences in the SDF results between the laboratories is a long and variable filtration time, both when the IDF fraction is filtered away from the TDF fraction and when the precipitated SDF fraction is separated in the second filtration. It is the experience of several laboratories that these problems can be overcome by using 0.5g or 0.25g test portions, when analyzing materials with levels of viscous fiber that hinder the filtration. Since preliminary results from an earlier study indicated that TDF analysis done independently yields the same results when compared with TDF

TABLE 3-7. Measures of Precision for Determining SDF.

Product	Average SDF, % dry wt	No. of Collaborators	Repeatability RSD_r, %	Reproducibility RSD_R, %
Cabbage	5.41	9	11.51	28.88
Carrots	11.02	11	11.37	15.76
French beans	10.85	9	4.71	8.75
Kidney beans	3.48	14	17.43	19.48
Butter beans	3.07	10	10.62	22.17
Okra	12.06	13	8.71	15.85
Onions	3.59	11	30.41	37.86
Parsley	5.13	9	21.83	56.86
Chick peas	1.35	12	39.06	44.38
Brussels sprouts	6.16	15	9.91	20.99
Barley	3.83	12	13.92	35.87
Rye Flour	3.35	15	13.96	19.25
Turnips	9.32	8	10.38	27.14
Soybran	7.08	10	14.66	14.66
Wheat germ	1.90	9	21.12	42.83
Seedless raisins	14.61	7	9.14	41.21
Mission figs	10.84	6	11.32	13.58
Calimyrna figs	18.15	5	5.10	15.39
Prunes	33.42	6	6.56	28.51
Apples	18.56	4	2.34	13.17
Peaches	27.30	6	3.09	11.61
Apricots	26.43	5	3.98	16.36

values obtained by summing IDF and SDF (Prosky et al. 1988), it is not unreasonable to determine SDF by the difference between TDF and IDF.

Tables A-1 through A-6 in the Appendix will contain values for TDF, SDF, and IDF, using the AOAC methods in a number of different laboratories.

Precision of Methods

In the 1985 collaborative study (which forms the basis for the current AOAC-TDF method), conducted under the auspices of the AOAC, nine laboratories in five countries analyzed nine different food matrices as blind duplicates. The results are summarized in Table 3-8. The average standard deviation for TDF in this study was 0.72%. The study in 1988, conducted by 13 laboratories in seven countries on ten food matrices (see Table 3-9) showed a standard deviation of 1.16%. In this same study, the standard deviation for insoluble dietary fiber was 1.25% and for soluble dietary fiber was 0.87%. It is obvious that the precision of the soluble dietary fiber method is not significantly different than that of either the total or insoluble dietary fiber methods.

TABLE 3-8. Standard Deviation of Total Dietary Fiber from AOAC Collaborative study (1985).

Sample matrix	Standard deviation (%)
Corn bran	1.36
Oats	0.58
Potatoes	0.54
Rice	0.56
Rye bread	0.35
Soy isolate	0.94
Wheat bran	1.13
Whole wheat flour	0.74
White wheat flour	0.27
Mean	0.72

To expand the scope of the methodology further, the 1989 AOAC collaborative study was carried out, this time on 22 additional food matrices of wide-ranging properties. The results are summarized in Table 3-10 (values have been corrected to whole food products, to take into account the dewatering and desugaring steps used in preparing the samples for collaborative study). As can be seen, the method had an average standard deviation of 0.84% for IDF and 0.53% for SDF.

Also in 1989, the American Association of Cereal Chemists, in its deliberations on a suitable definition for oat bran, conducted a collaborative study on the TDF, SDF, and IDF methods on oat groats, oat bran, oat

TABLE 3-9. Standard Deviation of Total, Insoluble, and Soluble Dietary Fiber (AOAC Collaborative study 1988).

Sample matrix	Standard deviation (%)		
	TDF	IDF	SDF
Corn bran	1.41	0.76	0.31
Fabulous Fiber	1.88	4.45	4.00
High fiber cereal	1.28	1.30	0.65
Oats	1.62	0.68	1.13
Potatoes	0.94	0.58	0.60
Rice	0.71	0.31	0.24
Rye bread	0.81	0.85	0.44
Soy isolate	1.38	1.80	0.46
Wheat bran	1.18	1.41	0.57
White wheat flour	0.42	0.33	0.27
Mean	1.16	1.25	0.87

TABLE 3-10. Standard Deviation of Insoluble and Soluble Dietary Fiber (AOAC Collaborative study 1989).

Sample matrix	Standard deviation (%)	
	IDF	SDF
Apples	0.10	0.09
Apricots	0.19	0.25
Barley	0.55	1.21
Beans, butter	1.72	0.60
Beans, French	1.34	0.85
Beans, kidney	0.92	0.60
Brussels sprouts	0.33	0.18
Cabbage	0.13	0.12
Carrots	0.26	0.13
Chick peas	2.48	0.53
Figs, Calimyrna	0.88	0.29
Figs, Mission	0.45	0.15
Okra	0.34	0.20
Onions	0.16	0.14
Parsley	0.55	0.34
Peaches	0.09	0.38
Prunes	1.96	2.08
Raisins	1.86	1.17
Rye flour	0.87	0.55
Soy bran	2.18	0.95
Turnips	0.28	0.20
Wheat germ	0.85	0.72
Mean	0.84	0.53

flour, and oat cereal. The results are shown in Table 3-11. The standard deviation for the SDF, IDF, and TDF methods are 1.02%, 1.77%, and 1.19%, respectively. Again, the precision of the SDF method is equivalent to that of the IDF method and better than that of the TDF method. Therefore, there is no advantage to be gained by directly determining total and insoluble dietary fiber and calculating soluble fiber by difference, instead of measuring IDF and SDF directly and determining the TDF by summation.

In a further collaborative effort, Lee et al. (1991) finished her study under the auspices of AOAC and AACC, to investigate the effect of a number of method changes on the precision of the method. These method changes have been approved by the AACC, and the results have been submitted for AOAC consideration. These results are summarized in Table 3-12. The standard deviation for IDF is 0.39%, for SDF is 0.39%, and for

TABLE 3-11. Standard Deviation of Total, Insoluble, and Soluble Dietary Fiber (AACC Collaborative study 1989).

Sample matrix	Standard deviation (%)		
	TDF	IDF	SDF
Oat groats (n=4)	1.14	1.86	1.14
Oat bran (n=4)	1.23	2.05	0.98
Oat flour (n=1)	1.16	1.02	1.11
Oat cereal (n=1)	1.21	1.05	0.58
Mean	1.19	1.77	1.02

n = number of samples of this matrix in collaborative

TDF is 0.59%. Again, the precision of the soluble dietary fiber method is better than that of the TDF method.

A summary of all results is shown in Table 3-13. The collaborative studies conducted under the auspices of the AOAC and AACC clearly show that the method for SDF is not significantly different in precision than the methods for IDF or TDF. In fact, in the 1988 and 1989 studies, the SDF method showed better precision than the IDF method. Therefore, there is no advantage to be gained by directly determining TDF and IDF and calculating SDF by difference, instead of measuring IDF and SDF directly and determining TDF by summation. The former procedure may in fact result in poorer precision when quantitating each of the

TABLE 3-12. Standard Deviation of Total, Insoluble, and Soluble Dietary Fiber AACC/AOAC Collaborative Study (1990).

Sample matrix	Standard deviation (%)		
	TDF	IDF	SDF
Apricots	0.09	0.12	0.11
Barley	0.85	0.61	0.62
Carrots	0.13	0.16	0.18
Green beans	0.07	0.08	0.11
High fiber cereal	0.94	0.71	0.56
Oat bran	2.06	1.17	1.14
Parsley	0.11	0.19	0.10
Prunes	0.40	0.09	0.31
Raisins	0.16	0.07	0.16
Soy bran	1.07	0.73	0.60
Mean	0.59	0.39	0.39

TABLE 3-13. Summary Standard Deviation of Total, Insoluble, and Soluble Dietary Fiber AACC/AOAC Collaborative Studies.

Study	Matrices	Standard deviation (%)		
		TDF	IDF	SDF
AOAC 1985	9	0.72	—	—
AOAC 1988	10	1.16	1.25	0.87
AOAC 1989	22	—	0.84	0.53
AACC 1989	10	1.19	1.77	1.02
AACC/AOAC (1990)	10	0.59	0.39	0.39
Mean		0.92	1.01	0.66

entities. To use a hypothetical situation, for example: If an analysis gives a TDF of 30 +/- 0.92%, and an IDF of 20 +/- 1.01%, the calculated SDF would have a value of 10 +/- 1.37%. On the other hand, direct assay for SDF and IDF would give an IDF of 20 +/- 1.01%, an SDF of 10 +/- 0.66%, and a calculated TDF of 30 +/- 1.21%, a definite improvement in the precision of the SDF, at the expense of some small loss of precision in the TDF.

MARKETING

Public acceptance of the hypothesis that dietary fiber has several major health benefits has opened up opportunities for the food technologist in promoting already existing fiber-rich foods and the development of new high-fiber foods and new high-fiber additives as supplements for foods.

Removing the fiber from the agricultural commodity by processing into an ingredient in preparation for sale through the marketplace has generally resulted, in the past, in ingredients that produce products with greater consumer appeal than products produced from those commodities with their fiber intact. Over the decades, the overall appeal in terms of the appearance, taste, texture, and ease of preparation of foods made with these ingredients has been a driving force in their increased production and consumption.

Increasing the level of dietary fiber in the diet of the general population offers a substantial challenge and a unique opportunity for the food technologist of today and in the near future. The challenge (and the commensurate opportunity) is to formulate and produce foods that contain high levels of dietary fiber that have a customer appeal equal to or greater than the customer appeal of the foods currently on the market. Health education programs and the resulting increase in health consciousness on the part of consumers will result in a significant permanent change in the eating habits of some (albeit a small) portion of the population. In addition, another significant portion of the population will temporarily change the makeup of their diet in an attempt to pursue healthier nutritional alternatives. However, these individuals attempting to change their consumption patterns will be tempted to revert back to old familiar eating habits unless there is an adequate supply of high dietary fiber products that are visually attractive, have appealing mouthfeel and texture, provide exciting, appetizing, and satisfying flavors, and are either ready-to-eat or easy to prepare. In all likelihood, the single most important factor in implementing significant long-term changes in total dietary fiber intake in the general population will be the ingenuity of the food technologist. Through the successful incorporation of high-fiber ingredients into appealing new products by formulation and processing changes, overall consumption patterns for dietary fiber will shift. Even small changes in fiber level in foods traditionally consumed in large quantity can result in a significant relative change in intake.

At the disposal of the food technologist is a whole host of both traditional and new ingredients, for use in pursuing potential new products. Many of the new ingredients are age-old, familiar ones that have been modified functionally in some way, to aid in producing products with improved visual, taste, textural, and preparation appeal, while still providing an enhanced level of dietary fiber. The ingredients discussed on the following pages represent some of the many currently available options. The discussion is by no means exhaustive; rather it is intended to demonstrate the wide variety of available options at the current time. With the many benefits of increased fiber in the diet, discussed in the first chapter, and with the expectation that the ongoing research in this area will reveal even more benefits, there will likely be a significant expansion of available ingredient options in the future.

Nutritional labeling that provides the consumer with up-to-date information regarding dietary fiber intake will be an important part of the whole process of increasing the level of dietary fiber consumed. The leaders in the area of producing enhanced dietary fiber products and providing information (via nutritional labels) have been the producers in the cereal food industry. As the research on dietary fiber and its effects progressed, and the benefits of increased dietary fiber in the diet became evident, the cereal foods industry (which traditionally had offered many whole grain-based products) began producing additional products, incorporating beneficial high-fiber ingredients. At the same time, information regarding the levels of dietary fiber in the food was being transmitted to the consumer by means of expanded information on the nutritional label. The industry as a whole has been a traditional supporter of complete, meaningful, and appropriate nutritional labeling and commensurate regulation on all important nutritional aspects of food products. Cereal industry support on labeling of dietary fiber has not been an exception to that traditional support.

The fiber products mentioned in this chapter are for informational purposes only, and the mention of a product in no way means an endorsement of the product by the authors.

Ingredients

Quality ingredients having the functionalities desired for incorporation into appealing, enhanced fiber food products are the mainstay to new product development, in the quest to increase dietary fiber intake in general. Many of the advanced technologies that led to the development of low-fiber ingredients and subsequent products with widespread consumer appeal can be applied to produce high-fiber alternatives. The following is a discussion of just some of the ingredients available at this time, with many more expected in the future, as market forces increase the demand. A regular perusal of the professional and trade magazines directed at the food technologist, such as Bakery Production and Marketing, Cereal Foods World, Food Business, Food Engineering, Food Processing, Food Technology, International Food Engineering, Milling and Baking News, Prepared Foods, Reference Source (for wholesale baking), and many others, will uncover a wealth of additional information on these and many other ingredients.

With the wide range of dietary fiber sources, types, and functionalities available, numerous opportunities present themselves for initiating dietary changes among consumers, by way of making available appealing products. To the extent that saturated fat, often shown as a risk factor in many of the diseases discussed in Chapter 1, can be displaced with dietary fiber, a twofold dietary improvement will be attained. Simply reducing the fat level of products by replacing the fat with emulsifiers, themselves often saturated fat-based or primarily saturated fatty acids, does little in the way of product improvement, in this regard. On the other hand, an ingredient or a combination of ingredients high in dietary fiber that allow a replacement or a significant reduction in ingredients that add saturated fat to the product should help promote the overall health of the consumers of that product.

Whole Food Sources

Whole food sources make up the most widely available and plentiful supply of ingredients providing a significant quantity of dietary fiber. In fact, these sources are in such a large supply that they are bought, sold, and traded as commodity items for the most part. Most of them have unique flavor, color, or texturizing properties that lend themselves to products of which they are the major ingredient. In some cases, they also lend themselves to inclusion in products as a minor ingredient that will increase the level of dietary fiber, without unduly changing the characteristics of the major components or of the product itself. Wheat, oats, and corn (as their whole grain flours), and fruits and vegetables (supplied and eaten fresh or canned, with minimal loss of dietary fiber) are the most widely used whole food items. Popcorn, obviously a whole grain item, can be consumed with minimal processing or additional ingredients, or it can be incorporated into a wide variety of products, particularly snack items. Other ingredients available as wholemeal flours include amaranth, barley, millet, rice, and rye.

As mentioned earlier, fresh or canned fruits and vegetables that are essentially intact can provide significant quantities of fiber to the diet. In addition, concentrated sources of these fibers can be obtained through dehydration processes. Heated forced air dehydration results in removal of the water, accompanied with cell structure collapse, which may affect the appearance and texture of the product, but fiber content and efficacy should remain essentially unchanged. For improved ingredient appearance and texture, freeze dried dehydration removes the water, while leaving the cell wall intact. California Vegetable Concentrates have available a wide variety of dehydrated vegetables and fruits that can potentially be used as ingredients for high-fiber products, with enhanced visual and flavor appeal. Gourmet Club Corporation (NJ) offers a similar line of products. Sweetened dried cranberries (called Craisins™, made by Ocean Spray (MA), are reported to have 6.2% to 9.9% total dietary fiber and are available as potential ingredients in breads, muffins, bagels, doughnuts, and pies. Ocean Spray also promotes a line of fruit powders, including raspberry, cranberry, and strawberry that contain 90+% fruit solids. Raisins, of course, are one of the oldest dehydrated fruits that are used in other products. The California Raisin Advisory Board claims that California raisins contain 5.3% total dietary fiber, of which nearly 60% is soluble dietary fiber. Dole Dried Fruit and Nut company offers dates, prunes, and raisins as ingredients. Fresh sliced or diced apples, processed on a daily basis, are promoted by Baker Apple Products Inc. (NY). Dried apples can be obtained from Vacu-Dry (CA). Concentrated apple fiber is available from TreeTop (WA).

Concentrated Fiber Sources

With the exception of whole grains, cereal brans are probably the most readily available sources of cereal fiber. During the milling process, the fiber is concentrated in the bran fraction; therefore, the food technologist wishing to achieve high levels of fiber in a formulated product needs almost certainly to consider bran in the formulation. In the past, bran additions to a formula were generally considered to impart negative properties to the product in terms of appearance, texture, and taste. Depending upon the end use, this can still be the case, but numerous ingredient suppliers are diligently pursuing processes that will overcome these hurdles, the objective being to provide bran ingredients that produce products not only as good as, but better than, their low-fiber counterparts.

Currently, the most popular fiber-rich food products are those that contain oat bran fiber. Oat bran fiber finds its way into reduced-calorie breads, cereal products, baked goods, beverages, and meat substitutes. One such product advertises that because of oat bran fiber's water-absorbing capacity in reduced-calorie breads, you can use less fiber and more water, thereby cutting the cost of a loaf of bread substantially. This Advanced Oat Fiber™, made by Williamson Food Ingredients (KY), makes the bread more palatable, while requiring less additional vital wheat gluten in the product. The total dietary fiber of the product is claimed to be 98% (dry weight basis), 78% insoluble, and 20% soluble and provides a very smooth mouth feel.

Snowite™ Oat Fiber from Canadian Harvest (MN) is promoted as a noncaloric, 90% total dietary fiber product that is bland in taste and light in color. National Oat Company (IA) produces a full line of milled oat products, including those high in oat fiber. Stabilized oat bran is also supplied by Wilke International Inc. (KS).

Oat hulls make up 30% of the weight of the oat kernel and are currently used as animal feed or for producing industrial solvents. The fiber composition of these hulls is about 50% hemicellulose, 40% crude cellulose, and 10% lignin (Webster 1986). Dougherty et al. (1988) presented formulations developed for reduced calorie bread, three varieties of soft-textured cookies, and a shell pasta, using oat hull and other dietary fiber sources as noncaloric ingredients. In summary, they were able to produce these products without the gritty mouthfeel problems encountered by other researchers (Pollizoto et al. 1983; Schimberni et al. 1982). This was probably due to improved processing in the conversion of oat hulls to oat fiber. The bleaching process, which improved the color and flavor, made the product comparable to standard products already on the grocery shelf, despite the decreased calories and increased dietary fiber. They did show that regular oat fiber, as well as bleached oat fiber, can be used successfully as an ingredient in the products. The bleaching process incorporates H_2O_2 as a bleaching agent (instead of chlorine-based bleaching compounds) and, as such, may result in the total absence of dioxin-like compounds in the final bleached products (Kehoe et al. 1990).

The properties of rice bran as a food stuff are the subject of an excellent article by Saunders in Cereal Foods World (1990). In it, he discusses harvesting the rice and drying and milling, to remove the hull, yielding brown rice. In the next step, the outer brown layer of the brown rice is removed by abrasive milling, yielding the familiar white rice. The separated brown layer is called rice bran, which, in the United States, includes the germ.

In the case of parboiled rice bran, the field rice is subjected to soaking and steaming before being dried and milled. The hull is first removed, as above, followed by the bran, to yield parboiled white rice and bran. Parboiled rice bran contains substantially less starch than bran from rice not parboiled. It also has a higher level of other nutrients. In the case of both parboiled and regular bran, they are subjected to a short-term, high temperature heat treatment immediately after milling, to destroy lipase activity and make the stabilized bran that is to be used in foods.

The total dietary fiber and soluble dietary fiber of parboiled and stabilized rice bran varies with the degree of milling and the amount of starch present in the bran (Table 3-14).

Rice bran can be added to baked goods, breads, snacks, and extruded food stuff.

Rice bran may be as effective as oat bran in lowering blood serum cholesterol (Duxbury 1989; Haumann 1989; Rutger 1990; Kahlon et al. 1989; Gerhardt and Gallo 1989; Hegsted et al. 1990), and this stimulus has led to many new rice products entering the market. Although there has been little published regarding the incorporation of rice bran into yeast breads, several manufacturers are producing breads containing rice bran. Substituting rice bran for wheat flour in yeast breads reduces the

TABLE 3-14. Dietary Fiber Contents of Stabilized and Parboiled Rice Bran (Dry Basis).

Bran type	Total Dietary Fiber, % dry wt	Soluble Dietary Fiber, % dry wt
Stabilized bran	20–25	1.8–2.6
Parboiled bran	31–33	2.0–2.5
Defatted stabilized bran	24–28	2.0–2.4
Defatted parboiled bran	44–51	2.4–2.9

From Saunders, R.M. *Cereal Foods World* (1990) 35:634.

gluten content of the formulation and may decrease the volume of the loaf. When Skurray et al. (1986) replaced 15% of the wheat flour with parboiled rice bran or fat-extracted rice bran, they noted an 8.7% and 22.9% decrease in volume of the loaves, respectively. They also noted no significant differences in the sensory values of the bread containing the parboiled rice bran.

Since water and fat absorption, foaming capacity, and stability of the rice bran are affected by the amount of fat in the rice bran (James and Sloan 1984) and because there appears to be considerable variation in the attributes of rice bran from different suppliers (James et al. 1983), the study by Sharp and Kitchens (1990) was designed to examine the influence of 1) substituting a commercially milled rice bran for wheat flour and 2) adding additional rice flour to the bran, to simulate an excessively milled product. The data suggest that rice bran can be substituted successfully for 15% of the wheat flour in yeast bread, without affecting loaf size, weight, or volume. When 30% of the wheat flour was substituted by rice bran, there was a significant reduction in volume. This indicates that many manipulations can be carried out with the basic formula, without sacrificing loaf volume. Keeping the loaf volume the same as the standard bread is an important prerequisite when making additions or substitutions in the bread. This is a very interesting point because Englyst had suggested that if resistant starch were included in the AOAC definition of dietary fiber, bread manufacturers could either heat the starch or bake the bread for an excessively long time, converting the enzymatically-digestible starch to resistant starch. This would then permit the manufacturer to declare a larger amount of dietary fiber in their product. The part of the question concerning the dietary fiber declaration has been answered earlier, but the question of who would buy the very small one-pound loaf of bread was not answered. Also, the bread would be hard from excessive baking.

Riceland Foods Inc. (AR) claims that its rice bran has 36% more dietary fiber than oat bran, and, due to their processing procedures, the bran has a sweet, toasted, nutty aroma and flavor and is never bitter. According to Farmers Rice Cooperative (CA), the natural sweetness and slightly nutty flavor of their FibeRice™ rice bran often carries through to the final product and thereby enhances the flavor of the prepared food. Calbran (California Rice Bran Inc.) has available specialty rice bran products, namely defatted rice bran, low-fat rice bran, brown rice flour, and textured rice bran. Pacific Rice Products Inc. offers their Fibran rice bran as either a high-fiber crisp rice or as a high-fiber nutty rice.

Solka-floc™, made by James River Corporation (NJ), is a powdered cellulose that has been in use for a long time but recently has found its way into food, where a reduction in fat content is desired. Normally, reducing the fat results in a loss of the unique texture and mouthfeel. This powdered cellulose counters that effect by providing solids and water-absorption properties, restoring the product back to its original characteristics. Solka-floc, which contains at least 90% cellulose (the remainder is hemicellulose) is promoted for reducing calories in food products by replacing lost bulk, while maintaining the texture, structure, and mouthfeel of the product. Avicel™ FD-100 cellulose gel, by FMC Corporation (PA), is promoted for use as a fiber source, with desirable functional properties in low-fat and nonfat food systems. JustFiber™ is a noncaloric, white, fat-free, flavor-free, and odorless source of alpha cellulose, from International Filler Corporation (NY).

The cellulose derivative carboxymethyl cellulose (Ticalose™ CMC by TIC Gums [MD]) is resistant to the effects of acid and salt, leading to superior stability in baked goods and sauces. CMCs are also claimed to be able to withstand extreme processing conditions, such as extended heating and high shear processing. Methyl cellulose (Methocel by Dow (MI); Benecel™ MC by Aqualon (DE)) are strong film formers and, as such, can serve as binders in batters or food matrices. Heating the gel makes it stronger. Since a methyl cellulose film serves as an oil barrier, it may not only add dietary fiber to the product, but reduce the fat content of the product as well, if the product is fried. Hydroxypropyl methylcellulose Benecel™ (HPMC by Aqualon [DE]) functions in a similar way, to help products retain moisture while reducing oil absorption.

Fibrex™ by Delta Fibre Food (MN) is a new fiber product that is derived from sugar beets. It is reported to have a total dietary fiber content of 74% and a soluble dietary fiber content of 24%. The fact that it is low in sugar and fat and free of gluten, starch, and phytic acid potentially makes it a good addition to cereals, baked products, pasta, soups, processed meats, and so on. The fact that it can hold five times its weight in water may also be an aid in improving the shelf life of the product it is added to. Duofiber by American Crystal Sugar (MN) is also reported to impart similar functional properties using this natural source of soluble and insoluble dietary fiber.

White wheat fiber by Watson Co. (CT) is a fiber product without odor or undesirable flavor. It is claimed to contain 98% total dietary fiber. Being white, bland, without aroma, and having superior water absorption properties make it an excellent potential additive to products, without alteration occurring in the unique properties of the product. Cerelife™ wheat fiber from Ceretech International Inc. (WA) is a light-colored wheat fiber and protein product reported to have 40% dietary fiber, 22% protein, 260 mg/100 g calcium, and 500 mg/100 g potassium. It is promoted for use in baked goods, batters, confections, dairy products, and soups. Quaker Toasted Wheat Bran, made by Quaker Oats Company, Industrial Chemicals Department (IL), is promoted for use in breads and baked goods for adding fiber and a toasty taste. Lauhoff Grain Company (IL), Illinois Cereal Mills, and Midwest Grain Products Inc (KS) also market wheat brans.

Canadian Harvest USA (MN) offers a complete variety of dietary fiber products that can find their way into baked goods, cereals, beverages, sauces, pasta, pharmaceuticals, and so forth. These products include oat fiber, white wheat bran, red wheat bran, full-fat wheat germ, toasted oat bran, stabilized oat fiber, and stabilized corn bran. Many of these types of products are also offered for sale by a variety of companies (Supreme Rice Mills, LA; Corn Products, IL, etc.).

A newer cereal fiber source, barley bran flour, has been evaluated as a dietary fiber ingredient in wheat bread. It was also compared for composition and baking performance to other fiber products on the market, such as oat, corn, wheat, and soy brans and cellulose and whole wheat flour (Chaudhary and Weber 1990). There were significant differences in both of these measurements when 15% of the wheat flour was replaced with the individual fiber ingredients. Barley bran flour outperformed other dietary fiber ingredients by producing a bread with substantially higher dietary fiber, highest loaf volume, and highest quality score for the enriched breads. Barley bran flour also scored the highest for flavor.

Barley, which is high in total and soluble dietary fiber at levels comparable to oats, has, in some of the hull-less varieties, a soluble dietary fiber equal to oat bran (Newman et al. 1988). Flours from four barley cultivars were compared with a wheat bread flour in terms of total and soluble dietary fiber, beta-glucan content, proximate analysis, and for color, volume, texture, and sensory acceptability of muffins baked from each (Newman et al. 1990). They reported that barley flour that was substituted 100% for the whole wheat flour provided a high-fiber muffin without adding bran. With the average muffin containing about 20 g of flour, any one of the barley muffins had about 7% or 1.4 g total dietary fiber compared to approximately 1.1 g dietary fiber, when whole wheat flour was used. An alternative procedure might be to substitute only part of the wheat flour with barley flour and substitute the remainder with a bran or other supplement. What this study showed was that acceptable muffins can be made with barley flour.

Barley fiber is available as a coarse flour-like ingredient from Minnesota Grain Pearling Company. The product is claimed to have a minimum of 50% total dietary fiber, with roughly 3% of the fiber being soluble dietary fiber. Miller Brewing Co. (WI) also supplies barley fiber ingredients under the Barley's Best tradename.

The authors would be remiss not to mention several of the companies that donated samples for the AOAC collaborative study. Vacu-Dry of California supplied dry, free-flowing forms of prunes, peaches, apricots, and apple solids for dietary fiber analysis. The results of the analysis of these samples are reported in the collaborative study of 1988 and were discussed earlier. Vitamins, Inc., of Illinois, supplied a defatted wheat germ for the same collaborative study. This product contained approximately 20% total dietary fiber and was prepared from freshly milled wheat germ, with most of the fat removed by solvent extraction. This stabilized product may be used as a nutritional ingredient in baked products, breakfast cereals, snacks, and dietetic foods, being available in granular, flour, and extruded forms.

Raisins and figs were generously supplied by the California Raisin Advisory Board and the California Fig Advisory Board, respectively, both of California. When these products were dried and sugar extracted, they were among the highest in total dietary fiber of any of the foods analyzed, both having about 65% total dietary fiber on a dry, sugar-free, weight basis.

Soy bran donated by Solnuts, Inc. (IA) had a total dietary fiber content of 72%. This product could also be used in a number of different foods for calorie reduction and high-fiber content, with the addition depending on the technology required to arrive at a final product.

Sunfiber is a new food ingredient produced by modifying guar gum, using enzymatic hydrolysis to reduce its molecular weight. Technologically, this treatment reduces the viscosity of the ingredient, while still retaining many of the beneficial physiological effects associated with guar gum. The applications for this ingredient are still being explored.

Inulin, the most abundant nonstructural polysaccharide in plants, is also not digested by human intestinal enzymes. It thus offers interesting nutritional properties as a carbohydrate replacement in low-calorie foods. While inulin-rich ingredient sources, such as Jerusalem artichokes, have been incorporated into pasta products, other potential uses of the materials remain to be explored and expanded, before they will have impact in the marketplace.

Carrageenan can be used in low levels, as an effective binder that prevents the separation of fat and water, offering the formulator a wider range of ingredient combination and level options, according to Carrageenan Marketing Corporation (CA), regarding its Carrageenan Bengel MB series.

Konjac flour (Nutricol™ by FMC Corporation [PA]) is claimed to have a minimum of 55% indigestible carbohydrates and is said to duplicate the functionality of egg whites, particularly in pasta products.

Pea fiber is available as a specialty fiber ingredient from Ogilvie Mills Inc. (MN).

All fiber-rich products that come to the marketplace may be used as foods, as long as they are naturally occurring or minimally processed. It stands to reason that an ingested dietary fiber should not be present in a food in such a great amount that it might be considered harmful. For example, the extract of a fiber from 100 pounds of a particular vegetable could be formulated into an ingredient that could be added to one or two slices of bread. Since the person would have to eat 100 pounds of the vegetable to get the dietary intake he would be ingesting in one or two loaves of bread, the additive would certainly have to undergo close scrutiny by the U.S. FDA for safety considerations. On the other hand, limited safety testing for an ingredient or food product that has had some unique preparation, aside from the usual milling, drying, and processing steps, will be required. For example, such a product could be a fiber extract that has been bleached by using an unusual chemical, or bacteriologically treated to hydrolyze the molecule, or treated in any manner so much so that essentially a new product is produced that in no way resembles the original one. One would have to show that all of the bleaching agents or bacteria are removed from the final product and that these treatments are indeed acceptable to the FDA.

All other products and ingredients purported to behave like dietary fiber would require a host of additional tests, to show that the product was safe at the intended level and usage, made a nutritional contribution to the diet, and is dietary fiber by the analysis procedures described by the AOAC. Of course, should any medical claim for the product be made, the company would have to submit a new drug application to the FDA, which would entail the usual lengthy testing procedures.

Summary

We have discussed a great deal about soluble and insoluble dietary fiber, and we have alluded to "resistant starch" and the foods that contain these components. To summarize, soluble dietary fiber is the water-soluble material that is not digestible by appropriately chosen enzymes, mimicking the human alimentary system, which is reprecipitatable when the water it is dissolved in is mixed with four parts of alcohol. It is less abundant in foods than insoluble dietary fiber. Soluble fibers are basically food gums from a variety of sources: beta glucans from grains (particularly oats and barley), pectins from fruits and some vegetables, agar, alginate, and carrageenan from seaweed, modified celluloses, and other gums, such as arabic, flax seed, ghatti, guar, karaya, locust bean, tragacanth, and xanthan. Insufficient consumption of soluble dietary fiber or increased consumption of soluble dietary fiber has been shown to affect a number of health states. Soluble dietary fiber consumption relates to a short-term insulin demand in diabetics and an increase in long-term peripheral insulin sensitivity. Decreases in hypertension have been related to soluble dietary fiber consumption, as have decreases in body weight. Satiety increases with an increased level of soluble dietary fiber intake. Soluble dietary fiber supplementation of the diet can result in significant reductions in serum cholesterol, of low density lipoprotein (LDL) associated cholesterol, and an improved ratio of high density lipoprotein (HDL) associated cholesterol to LDL cholesterol. All of the above improvements in health state in turn relate to a significant decrease in the risk of coronary heart disease, for the individual who is able to shift these factors as a result of dietary changes, resulting in higher intakes of soluble dietary fiber. Soluble fiber makes up roughly 1/3 of the fiber in the food supply. If, as discussed earlier, individual dietary fiber consumption should be 30 grams per day (12 grams per 1000 kilocalories), the daily intake of soluble dietary fiber should be 4 grams per 1000 kilocalories.

Insoluble dietary fiber, the food material that is not water-soluble and is not digestible by appropriately chosen enzymes mimicking the human alimentary system, is the major component of dietary fiber in foods. Insoluble dietary fiber consists primarily of cellulose, with lesser amounts of hemicellulose, lignin, cutin, plant waxes, and sometimes insoluble pectins. Like soluble dietary fiber, insoluble dietary fiber may have a positive effect on diabetes, as evidenced by the decrease in the disease state during the years when inhabitants of England and Wales had to consume low-extraction flours (with the resulting increase in dietary fiber). Insoluble dietary fiber may also play a role in reducing the risk of coronary heart disease and obesity, even if only by virtue of the inverse relationship between fiber intake and fat intake and the subsequent reduction in calorie intake. Insoluble dietary fiber plays a major role in the function of the alimentary processes, helping the individual to improved physical and psychological health, as a result of normal bowel function. In particular, increased levels of insoluble dietary fiber consumption cause reduced rates of constipation, hemorrhoids, colon cancer, appendicitis (perhaps), Crohn's disease (including ileitis), diverticular disease, irritable bowel syndrome (also called colitis), and varicose veins. A recommended consumption level of 12 grams of total dietary fiber per 1000 kilocalories would mean a consumption level of approximately 8 grams of insoluble dietary fiber per 1000 kilocalories (20 grams per day).

Although the makeup, effects, and extent of occurrence in foods of "resistant starch" has not as yet been fully elucidated, a number of characteristics of this material have been studied sufficiently to warrant its consideration as a portion of the total dietary fiber content of the food in which it is found. In many ways, it behaves similarly to soluble dietary fiber, passing undigested through the small intestine and degraded in the large intestine by bacterial flora. As pointed out earlier, it is not certain, at this point, what quantity of the food starch behaves this way in the intestinal tract of any given individual during any given eating occasion. Studies with ileostomy patients tend to indicate that a larger portion of the starch than is being isolated in the laboratory as nondigestible may be passing through the small intestine undigested. A number of current analytical methodologies utilized in quantitating dietary fiber clearly indicate that a portion of the starch in some foods clearly is not digestible by the amylases that normally digest and solubilize the digestible carbohydrate portion of the food. It is uncertain whether methodologies can be developed that will quantitate additional quantities of starch that may be undigested, until such time as it can be ascertained that the additional starch is consistently undigested at a reasonably constant level, regardless of differences in the diet or in the energy requirements of the individual. Current techniques to quantitate "resistant starch" utilize calculation of difference in quantitatable aspects of the sample, with and without treating the sample with either alkali or dimethyl sulfoxide. The extent to which either of these treatments affects the quantitation of nonstarch components of the dietary fiber is uncertain. Since the starch isolated as part of the currently accepted dietary fiber method is clearly nondigestible and behaves similarly to soluble dietary fiber, it should be included in the quantitation. At such future time when the extent of additional indigestible starch is clearly characterized, the analytical methods should be updated, as necessary, to include that additional quantity as well.

References

Asp, N.-G., C.-G. Johansson, H. Hallmer, and M. Siljeström. 1983. Rapid enzymatic assay of insoluble and soluble dietary fiber. *Journal of Agricultural and Food Chemistry* 31:476–482.

Chaudhary, V.K., and F.E. Weber. 1990. Barley bran flour evaluated as dietary fiber ingredient in wheat bread. *Cereal Foods World* 35:560–562.

Dougherty, M., R. Sombke, J. Irvine, and C.S. Rao. 1988. Oat fibers in low calorie breads, soft-type cookies, and pasta. *Cereal Foods World* 33:424-427.
Duxbury, D.C. 1989. Stabilized rice bran reduces cholesterol. *Food Processing* 50(8):106-109.
Gerhardt, A.L., and N.B. Gallo. 1989. Effect of a processed medium-grain rice bran and germ product on hypercholesterolemia. *American Association of Cereal Chemists* annual meeting, Washington, D.C.
Handbook of dietary fiber in human nutrition. 1986. ed. G.A. Spiller. Boca Raton, FL: CRC Press Incorporated.
Haumann, B.F. 1989. Rice bran linked to lower cholesterol. *Journal of the American Oil Chemists' Society* 66:615-618.
Hegsted, M., M.M. Windhauser, F.B. Lester, and S.K. Morris. 1990. Stabilized rice bran and oat bran lower cholesterol in humans. *Federation of American Societies for Experimental Biology Journal* 4:A368, Abstract no. 590.
James, C., S. Sloan, and D. Gadbury. 1983. Utilization of rice bran in baked goods. *Arkansas Farm Research* 32:11.
James, C., and S. Sloan. 1984. Functional properties of edible rice bran in model systems. *Journal of Food Science* 49:310-311.
Kahlon, T.S., R.M. Saunders, F.I. Chow, M.C. Chiu, and A.A. Betschart. 1989. Effect of rice bran and oat bran on plasma cholesterol in hamsters. *Cereal Foods World* 34:768.
Kehoe, D.F., L.L. Kong, P.J. Tisdel, K.H. MacKenzie, R.L. Hall, W.E. Field, and C.S. Rao. 1990. Subchronic safety study of bleached oat bull fiber in rats. *Cereal Foods World* 35:1026-1029.
Lee, S.C., L. Prosky, and J.W. DeVries. 1991. Determination of soluble/insoluble and total dietary fiber in foods: collaborative study. *Journal of the Association of Official Analytical Chemists.* In press.
Newman, R.K., C.F. McGuire, and C.W. Newman. 1990. Composition and muffin-baking characteristics of flours from four barley cultivars. *Cereal Foods World* 35:563-566.
Newman, R.K., C.W. Newman, and C.F. McGuire. 1988. Barley B-glucans in foods: functional properties and hypocholesterolemic effects. *Proceedings 38th Australian Cereal Chemists Conference—Chemistry in Australia* 55:316.
Nutrition labeling of foods; calorie content. 1987. *Federal Register* 52:28690-28691.
Pollizoto, L.M., A.M. Tinsley, C.W. Weber, and J.W. Berry. 1983. Dietary fibers in muffins. *Journal of Food Science* 48:111-113.
Prosky, L., N.-G. Asp, I. Furda, J.W. DeVries, T.F. Schweizer, and B.F. Harland. 1984. Determination of total dietary fiber in foods and food products, and total diets: interlaboratory study. *Journal of the Association of Official Analytical Chemists* 67:1044-1052.
Prosky, L., N.-G. Asp, I. Furda, J.W. DeVries, T.F. Schweizer, and B.F. Harland. 1985. Determination of total dietary fiber in foods and food products: collaborative study. *Journal of the Association of Official Analytical Chemists* 68:677-679.

Prosky, L., N.-G. Asp, T.F. Schweizer, J.W. DeVries, and I. Furda. 1988. Determination of insoluble, soluble, and total dietary fiber in foods and food products: interlaboratory study. *Journal of the Association of Official Analytical Chemists* 71:1017-1023.

Rutger, R.J. 1990. Rice and oat brans do lower cholesterol. *Rice Journal* 93(2):25.

Saunders, R.M. 1990. The properties of rice bran as a foodstuff. *Cereal Foods World* 35:632-636.

Schweizer, T.F., E. Walter, and P. Venetz. 1988. Collaborative study for the enzymatic, gravimetric determination of total dietary fibre in foods. *Mitteilungen aus dem Gebiete der Lebensmitteluntersuchung und Hygiene* 79:57-68.

Schimberni, M., F. Cardinali, G. Sodini, and M. Cannela. 1982. Chemical and functional characterization of corn bran, oat hull flour, and barley flour. *Lebensmittel-Wissenschaft und Technologie* 15:337-339.

Sharp, C.Q., and K.J. Kitchens. 1990. Using rice bran in yeast bread in a home baker. *Cereal Foods World* 35:1021-1024.

6th Annual Association of Official Analytical Chemists Spring Workshop. *Discussion of the definition and analysis of fibre.* 1981. Ottawa, Canada.

Skurray, G.R., D.A. Woolridge, and M. Nguyen. 1986. Rice bran as a source of dietary fiber in bread. *Journal of Food Technology* 21:727-730.

Southgate, D.A.T. 1986. The relation between composition and properties of dietary fiber and physiological effects. In *Dietary fiber basic and clinical aspects*, ed. G. Vahouny and D. Kritchevsky, pp. 35-48. New York and London: Plenum Press.

Total dietary fiber in foods, enzymatic method, first action. 1985. *Journal of the Association of Official Analytical Chemists* 68:399.

Total dietary fiber in foods, enzymatic gravimetric method, amended final action. 1986. *Journal of the Association of Official Analytical Chemists* 69:370.

Webster, F.H. ed. 1986. *Oats: chemistry and technology*. American Association of Cereal Chemists, St. Paul, MN.

Appendix

Tables A-1, A-2, A-3, A-4, A-5, and A-6 contain the values for the dietary fiber content of foods, determined by the AOAC method. They represent the values derived from analyzing foods and food products in numerous laboratories.

Table A-1 gives the total dietary fiber for more than 300 foods and their varieties of preparation, prior to eating. This provisional table was put out by the Human Nutrition Information Service at the U.S.D.A., by the capable work of R.H. Matthews and P.R. Pehrsson, in 1988.

Table A-2 shows the total dietary fiber content, as determined by the AOAC method, of more than 250 varieties of Japanese foods and preparations. It is of interest to note that the Japanese who consume large amounts of fish have reported dietary fiber values for these and other animal products not normally thought to contain significant amounts of fiber (the traditional definition of fiber is material that is only of plant cell origin). However, a potential definition of "edible fiber" from all sources has been considered (see Chapter 1). According to the Food and Agricultural Organization of the United Nation World Health Organization, dietary fibers are constituents of substances of both animal and plant origins that are resistant to hydrolysis by human alimentary enzymes. They are classified as 1) foods of plant origin, which includes the cell wall structural materials, such as cellulose, hemicellulose, and lignin, and the nonstructural materials (storage polysaccharides), such as pectin, guar gum, konjac mannan, alginic acid, and rhamnarin; 2) foods of animal origin, such as chitin, chondroitin sulfate, and collagen; and 3) polysaccharide derivatives, such as methylcellulose, carboxymethylcellulose, alginic acid propylene glycol ester, and chitosan. With this definition in mind, the Ministry of Health and Welfare of Japan has recently released official tables of the dietary content of major food items.

Table A-3 shows the total dietary fiber content of over 100 baked products and cereals, determined by the AOAC method. This was the first paper to report analysis of a large number of diverse foods, using the official method.

Table A-4 shows the insoluble, soluble, and total dietary fiber of a variety of cooked and raw vegetables, as determined by the AOAC method when employed in the laboratories of C. Lintas in Italy.

Table A-5 shows the total dietary fiber content of fruits, breads, grains, cereals, rice, legumes, and vegetables analyzed by the AOAC method. Most of the values were derived from the Australian laboratories of Visser and Gurnsey, but values from six other laboratories have been incorporated into the table.

Finally, Table A-6 shows the soluble and total dietary fiber values of miscellaneous cereal grains and products and has been derived from data generated in eight laboratories.

TABLE A-1. Total Dietary Fiber Content of Foods by the AOAC Method*.

Food Item	Moisture	TDF
	g per 100g edible portion	
Baked Products		
Bagels, plain	31.6	2.1
Biscuit mix:		
Dry	8.7	1.3
Baked	29.4	1.8
Biscuits, made from refrigerated dough, baked	28.7	1.5
Breads:		
Boston brown	47.2	4.7
Bran	37.7	8.5
Cornbread mix:		
Dry	6.0	6.5
Baked	34.4	2.6
Cracked-wheat	35.9	5.3
French	33.9	2.3
Hollywood-type, light	37.8	4.8
Italian	34.1	2.7
Mixed-grain	38.2	6.3
Oatmeal	36.7	3.9
Pita:		
White	32.1	1.6
Whole-wheat	30.6	7.4
Pumpernickel	38.3	5.9
Reduced-calorie, high-fiber:		
Wheat	43.7	11.3
White	41.8	7.9
Rye	37.0	6.2
Vienna	—	3.2
Wheat	37.0	3.5
Toasted	—	5.2
White	37.1	1.9
Toasted	—	2.5
Whole-wheat	38.3	7.4
Toasted	—	8.9
Bread crumbs, plain or seasoned	5.7	4.2
Bread stuffing, flavored, from dry mix	65.1	2.9
Cake mix:		
Chocolate:		
Dry	3.8	2.4
Prepared	33.3	2.2

TABLE A-1. Continued.

Food Item	Moisture	TDF
	g per 100g edible portion	
Yellow:		
Dry	4.1	1.1
Prepared	40.0	0.8
Cakes:		
Boston cream pie	47.6	1.4
Coffeecake:		
Crumb topping	22.3	3.3
Fruit	31.7	2.5
Fruitcake, commercial	22.0	3.7
Gingerbread, from dry mix	38.5	2.9
Cheesecake:		
Commercial	44.6	2.1
From no-bake mix	44.4	1.9
Cookies:		
Brownies	12.6	2.2
With nuts	12.6	2.6
Butter	4.7	2.4
Chocolate chip	4.0	2.7
Chocolate sandwich	2.2	2.9
Fig bars	16.7	4.6
Fortune	8.0	1.6
Oatmeal	5.7	2.9
Oatmeal, soft-type	—	2.7
Peanut butter	6.7	1.8
Shortbread with pecans	3.3	1.8
Vanilla sandwich	2.1	1.5
Crackers:		
Cheese, sandwich with peanut butter filling	4.0	1.1
Crisp bread, rye	6.1	16.2
Graham	4.1	3.2
Honey	4.1	1.7
Matzo:		
Plain	6.1	2.9
Egg/onion	8.0	5.0
Melba toast:		
Plain	5.6	6.3
Rye	6.7	7.9
Wheat	6.1	7.4
Rye	7.2	15.8
Saltines	—	2.6
Snack-type	4.2	1.2
Wheat	3.2	5.5
Whole-wheat	2.7	10.4

TABLE A-1. Continued.

Food Item	Moisture	TDF
	g per 100g edible portion	
Croutons, plain or seasoned	5.6	4.7
Doughnuts:		
Cake	19.7	1.3
Yeast-leavened, glazed	26.7	2.2
English muffin, whole-wheat	45.7	6.7
French toast, commercial, ready-to-eat	48.1	3.1
Ice cream cones:		
Sugar, rolled-type	3.0	4.6
Wafer-type	5.3	4.1
Muffins, commercial:		
Blueberry	37.3	3.6
Oat bran	35.0	7.5
Pancake/waffle mix		
Regular:		
Dry	8.7	2.7
Prepared	50.4	1.4
Buckwheat, dry	9.1	2.3
Pastry, danish:		
Plain	19.3	1.3
Fruit	27.6	1.9
Pies, commercial:		
Apple	51.7	1.6
Cherry	46.2	0.8
Chocolate cream	43.5	2.0
Egg custard	46.5	1.6
Fruit and coconut	—	0.9
Pecan	19.8	3.5
Pumpkin	58.1	2.7
Rolls, dinner, egg	30.4	3.8
Taco shells	6.0	8.0
Toaster pastries	8.9	1.0
Tortillas:		
Corn	43.6	5.2
Flour, wheat	26.2	2.9
Waffles, commercial frozen, ready-to-eat	45.0	2.4
Breakfast Cereals, Ready-to-eat		
Bran, high-fiber	2.9	35.3
Extra fiber	—	45.9
Bran flakes	2.9	18.8
Bran flakes with raisins	8.3	13.4

TABLE A-1. Continued.

Food Item	Moisture	TDF
	g per 100g edible portion	
Corn flakes:		
Plain	2.8	2.0
Frosted or sugar-sparkled	1.9	2.2
Fiber cereal with fruit	—	14.8
Granola	3.3	10.5
Oat cereal	5.0	10.6
Oat flakes, fortified	3.1	3.0
Puffed wheat, sugar-coated	1.5	1.5
Rice, crispy	2.4	1.2
Wheat and malted barley:		
Flakes	3.4	6.8
Nuggets	3.2	6.5
With raisins	—	6.0
Wheat flakes	4.3	9.0
Cereal Grains		
Amaranth	9.8	15.2
Amaranth flour, whole-grain	10.4	10.2
Arrowroot flour	11.4	3.4
Barley	9.4	17.3
Barley, pearled, raw	10.1	15.6
Bulgur, dry	8.0	18.3
Corn bran, crude	4.7	84.6
Corn flour, whole-grain	10.9	13.4
Cornmeal:		
Whole-grain	10.3	11.0
Degermed	11.6	5.2
Cornstarch	8.3	0.9
Farina, regular or instant:		
Dry	10.6	2.7
Cooked	85.8	1.4
Hominy, canned	79.8	2.5
Millet, hulled, raw	—	8.5
Oat bran, raw	6.6	15.9
Oat flour	7.8	9.6
Oats, rolled or oatmeal dry	8.8	10.3
Rice, brown, long-grain:		
Raw	11.1	3.5
Cooked	73.1	1.7
Rice, white:		
Glutinous, raw	10.0	2.8
Long-grain, raw	11.6	1.0

TABLE A-1. Continued.

Food Item	Moisture	TDF
	g per 100g edible portion	
Parboiled:		
Dry	10.5	1.8
Cooked	—	0.5
Precooked or instant:		
Dry	8.1	1.6
Cooked	76.4	0.8
Medium-grain, raw	12.9	1.4
Rice bran, crude	6.1	21.7
Rice flour:		
Brown	12.0	4.6
White	11.9	2.4
Rye flour, medium or light	9.4	14.6
Semolina	12.7	3.9
Tapioca, pearl, dry	12.0	1.1
Triticale	10.5	18.1
Triticale flour, whole-grain	10.0	14.6
Wheat bran, crude	9.9	42.4
Wheat flour:		
White, all-purpose	11.8	2.7
Whole-grain	10.9	12.6
Wheat germ:		
Crude	11.1	15.0
Toasted	2.9	12.9
Wild rice, raw	7.8	5.2
Fruits and Fruit Products		
Apples, raw:		
With skin	83.9	2.2
Without skin	84.5	1.9
Apple juice, unsweetened	87.9	0.1
Applesauce:		
Sweetened	79.6	1.2
Unsweetened	88.4	1.5
Apricots, dried	31.1	7.8
Apricot nectar	84.9	0.6
Bananas, raw	74.3	1.6
Blueberries, raw	84.6	2.3
Cantaloup, raw	89.8	0.8
Figs, dried	28.4	9.3
Fruit cocktail, canned in heavy syrup, drained	—	1.5
Grapefruit, raw	90.9	0.6
Grapes, Thompson, seedless, raw	81.3	0.7

TABLE A-1. Continued.

Food Item	Moisture	TDF
	g per 100g edible portion	
Kiwifruit, raw	83.0	3.4
Nectarines, raw	86.3	1.6
Olives:		
Green	—	2.6
Ripe	—	3.0
Oranges, raw	86.8	2.4
Orange juice, frozen concentrate:		
Undiluted	57.8	0.8
Prepared	88.1	0.2
Peaches:		
Raw	87.7	1.6
Canned in juice, drained	—	1.0
Dried	31.8	8.2
Pears, raw	83.8	2.6
Pineapple:		
Raw	86.5	1.2
Canned in heavy syrup, chunks, drained	79.0	1.1
Prunes:		
Dried	32.4	7.2
Stewed	—	6.6
Prune juice	81.2	1.0
Raisins	15.4	5.3
Strawberries	91.6	2.6
Watermelon	91.5	0.4
Legumes, Nuts, and Seeds		
Almonds, oil-roasted	3.3	11.2
Baked beans, canned:		
Barbecue-style	—	5.8
Sweet or tomato sauce:		
Plain	72.6	7.7
With franks	69.3	6.9
With pork	71.7	5.5
Beans, Great Northern:		
Raw	10.7	40.0
Canned, drained	69.9	5.4
Cashews, oil-roasted	5.4	6.0
Chickpeas, canned, drained	68.2	5.8
Coconut, raw	47.0	9.0
Cowpeas (black-eyed peas):		
Raw	12.0	27.0
Cooked, drained	70.0	9.6

TABLE A-1. Continued.

Food Item	Moisture	TDF
	g per 100g edible portion	
Hazelnuts, oil-roasted	1.2	6.4
Lima beans:		
Raw	10.2	19.0
Cooked, drained	69.8	7.2
Miso	47.4	5.4
Mixed nuts, oil-roasted, with peanuts	—	9.0
Peanuts:		
Dry-roasted	1.6	8.0
Oil-roasted	2.0	8.8
Peanut butter:		
Chunky	1.1	6.6
Smooth	1.4	6.0
Pecans, dried	4.8	6.5
Pistachio nuts	3.9	10.8
Sunflower seeds, oil-roasted	2.6	6.8
Tahini	3.0	9.3
Tofu	84.6	1.2
Walnuts, dried:		
Black	4.4	5.0
English	3.6	4.8
Miscellaneous		
Beer, regular	92.3	0.5
Candy:		
Caramels, vanilla	7.6	1.2
Chocolate, milk	0.8	2.8
Sugar-coated discs	—	3.1
Carob powder, unsweetened	1.2	32.8
Chili powder	9.1	34.2
Chocolate, baking	0.7	15.4
Cocoa, baking	1.3	29.8
Cocoa mix, prepared	79.8	1.2
Curry powder	8.7	33.2
Gravy, beef, canned	89.1	0.4
Jelly, apple	32.3	0.6
Milk, chocolate	82.3	1.5
Pepper, black	9.4	25.0
Pie filling:		
Apple	74.9	1.0
Cherry	69.7	0.6
Preserves:		
Peach	32.4	0.7

TABLE A-1. Continued.

Food Item	Moisture	TDF
	g per 100g edible portion	
Strawberry	31.7	1.2
Soup, canned, condensed:		
Chicken with noodles or rice	86.5	0.6
Vegetable	84.9	1.3
Yeast, active, dry	6.8	31.6
Pasta		
Macaroni, protein-fortified, dry	10.2	4.3
Macaroni, tricolor, dry	9.8	4.3
Noodles, Chinese, chow mein	0.7	3.9
Noodles, egg, regular:		
Dry	9.7	2.7
Cooked	68.7	2.2
Noodles, Japanese, dry:		
Somen	9.2	4.3
Udon	8.7	5.4
Noodles, spinach, dry	8.5	6.8
Spaghetti and macaroni:		
Dry	10.5	2.4
Cooked	64.7	1.6
Spaghetti, dry:		
Spinach	8.7	10.6
Whole-wheat	7.1	11.8
Snacks		
Cheese-flavored, corn-based puffs or twists	—	1.0
Corn, toasted	—	6.9
Corn chips	—	4.4
Barbecue-flavored	—	5.2
Granola bars, crunchy:		
Chocolate chip	—	4.4
Cinnamon	—	5.0
Popcorn:		
Air-popped	—	15.1
Oil-popped	—	10.0
Potato chips	2.5	4.8
Flavored	—	4.5
Potato chips, formulated	1.6	3.6
Pretzels	—	2.8
Tortilla chips	—	6.5
Flavored	—	6.2

TABLE A-1. Continued.

Food Item	Moisture	TDF
	g per 100g edible portion	
Vegetables and Vegetable Products		
Artichokes, raw	84.4	5.2
Beans, snap:		
Raw	90.3	1.8
Canned:		
Drained solids	93.3	1.3
Solids and liquid	94.5	0.8
Beets, canned:		
Drained solids, sliced	91.0	1.7
Solids and liquids	91.3	1.1
Broccoli:		
Raw	90.7	2.8
Cooked	90.2	2.6
Brussels sprouts, boiled	87.3	4.3
Cabbage, Chinese:		
Raw	94.9	1.0
Cooked	95.4	1.6
Cabbage, red:		
Raw	91.6	2.0
Cooked	93.6	2.0
Cabbage, white, raw	91.5	2.4
Carrots:		
Raw	87.8	3.2
Canned, drained solids	93.0	1.5
Cauliflower:		
Raw	92.3	2.4
Cooked	92.5	2.2
Celery, raw	94.7	1.6
Chives	92.0	3.2
Corn, sweet:		
Raw	76.0	3.2
Cooked	69.6	3.7
Canned, brine pack:		
Drained solids	76.9	1.4
Solids and liquid	81.9	0.8
Cream style	78.7	1.2
Cucumbers, raw	96.0	1.0
Pared	—	0.5
Lettuce:		
Butterhead or iceberg	95.7	1.0
Romaine	94.9	1.7

TABLE A-1. Continued.

Food Item	Moisture	TDF
	g per 100g edible portion	
Mushrooms:		
Raw	91.8	1.3
Boiled	91.1	2.2
Onions, raw	90.1	1.6
Onions, spring, raw	91.9	2.4
Parsley	88.3	4.4
Peas, edible-podded:		
Raw	88.9	2.6
Cooked	88.9	2.8
Peas, sweet, canned:		
Drained solids	81.7	3.4
Solids and liquid	86.5	2.0
Peppers, sweet, raw	92.8	1.6
Pickles:		
Dill	93.8	1.2
Sweet	68.9	1.1
Potatoes:		
Raw:		
Flesh and skin	80.0	1.8
Flesh	79.0	1.6
Baked:		
Flesh	75.4	1.5
Skin	47.3	4.0
Boiled	77.0	1.5
French-fried, home-prepared from frozen	52.9	4.2
Spinach:		
Raw	91.6	2.6
Cooked	91.2	2.2
Squash:		
Summer:		
Raw	93.7	1.2
Cooked	93.7	1.4
Winter:		
Raw	88.7	1.8
Cooked	89.0	2.8
Sweet potatoes:		
Raw	72.8	3.0
Cooked	72.8	3.0
Canned, drained solids	72.5	1.8
Tomatoes, raw	94.0	1.3
Tomato products:		
Catsup	—	1.6
Paste	74.1	4.3

TABLE A-1. Continued.

Food Item	Moisture	TDF
	g per 100g edible portion	
Puree	87.3	2.3
Sauce	89.1	1.5
Turnip greens:		
Raw	91.1	2.4
Boiled	93.2	3.1
Turnips:		
Raw	91.9	1.8
Boiled	93.6	2.0
Vegetables, mixed, frozen, cooked	83.2	3.8
Water chestnuts, canned, drained solids	87.9	2.2
Watercress	95.1	2.3

*Provisional table of dietary fiber content of selected foods. U.S.D.A., Human Nutrition Information Service. R.H. Matthews and P.R. Pehrsson.

TABLE A-2. Total Dietary Fiber Content of Japanese Foods by the AOAC Method or Modified AOAC Method.

Food item	Moisture	TDF
	g per 100g edible portion	
Cereal		
Oatmeal	9.60	7.46
Barley, milled and pressed	11.34	5.26
Barley, milled and cut	10.24	5.25
Soft flour	11.89	2.12
White bread	2.60	2.55
Bread-type rolls	4.67	1.99
Rye bread	4.22	5.21
Fiber bread	9.53	3.96
Raisin bread	3.57	3.35
Soft rolls	2.29	1.83
Udon noodles, raw, wet	0.84	1.45
Udon noodles, boiled	3.30	0.99
Udon noodles, raw, dry	11.80	2.09
Somen-Hiyamugi noodles, raw, dry	11.50	2.08
Chinese lo mein noodles, raw, wet	4.25	1.48
Chinese lo mein noodles, boiled	3.05	1.08
Chinese lo mein noodles, steamed	1.40	0.97
Chinese cooked noodles flying-dried	5.80	2.43
Chinese cooked noodles hot air-dried	12.00	2.08
Macaroni and spaghetti, dry	11.20	2.72
Wheat germ	3.10	11.12
Bread crumbs	8.06	3.36
Brown rice grain	14.40	2.92
Half-milled rice, yield 95%-96%	12.87	2.27
Under-milled rice, yield 93%-94%	6.34	1.73
Well-milled rice	13.80	0.72
Mochi glutinous rice cake	2.21	0.33
Sekihan glutinous rice and azuki beans	3.00	1.56
Rice bran	8.73	22.20
Soba buckwheat noodles, boiled	2.66	1.62
Soba buckwheat noodles, raw, dry	11.00	4.74
Popcorn, popped	3.45	9.71
Cornflakes	2.40	2.89
1:1 mixture of hard and medium flour	6.11	2.44
Potatoes and Starches		
Devil's tongue, block type	3.96	1.67
Devil's tongue, noodle type	2.40	3.62
Sweet potatoes, raw	6.00	2.32
Satoimo dasheen, raw	2.20	2.20

TABLE A-2. Continued.

Food item	Moisture	TDF
	g per 100g edible portion	
Potatoes, raw	4.00	1.35
Potato chips, fried	4.21	3.38
Corn starch	12.48	0.30
Yam tuber, Ichoimo, raw	1.18	1.43
Nagaimo Chinese yam, raw	4.22	0.87
Starch noodles, potato starch mixed	4.70	1.11
Sugars and Sweeteners		
Brown sugar, lump	1.07	0.17
Honey	—	0.06
Confectionaries		
Milk chocolate	0.80	4.02
Bun with minced beef, bean jam, etc.	4.94	1.16
Potato tips and other snacks, mixed	4.20	3.44
Cookies, mixed	4.97	1.60
Japanese crackers, mixed	4.37	1.37
Sponge cake and other cakes, mixed	5.00	0.80
Japanese cakes and sweet pastes, mixed	5.43	3.47
Nuts and Seeds		
Cashew nuts, roasted	8.53	3.98
Ginkgo nuts, raw	1.52	0.62
Chestnuts, raw	9.90	3.71
Chestnuts, roasted	12.47	7.02
Sesame seeds, dry	24.60	15.37
Sesame seeds, roasted	5.80	11.58
Peanuts, dry	2.10	7.66
Peanuts, roasted	3.80	8.68
Peanut butter	4.83	5.88
Pulses		
Azuki beans	13.80	15.97
Kidney beans, whole, dry	14.76	19.76
Uzuramame kidney beans, cooked	4.75	6.86
Peas, whole, boiled	2.17	5.21
Broad beans, whole, dry	2.40	19.53
Otafukumame broad beans, cooked	5.57	5.63
Soybeans, dry	8.00	15.03
Soybeans, boiled	3.49	7.11

TABLE A-2. Continued.

Food item	Moisture	TDF
	g per 100g edible portion	
Soybeans, defatted, whole	8.60	15.96
Soybeans, roasted, ground	1.71	17.14
Soybean curd (Tofu), momen medium	5.26	0.62
Soybean curd (Tofu), kinukosi fine	1.53	0.35
Soybean curd, baked (Yaki-tofu)	5.00	0.91
Soybean curd, steamed (Namaage)	3.17	1.71
Soybean curd, fried (Aburaage)	6.48	1.60
Ganmodoki	8.33	2.37
Soybean curd, dry (Kori-tofu)	6.97	7.35
Itohiki-natto	2.30	9.60
Miso, sweet type	12.10	4.28
Miso, dark yellow type	8.00	6.42
Okara	2.53	9.42
Soy milk, reconstituted	5.24	0.26
Soybeans, cooked, Yuba mixed	6.21	5.65
Green peas, Adzuki beans, etc.	5.65	4.25
Fishes and Shellfish		
Horse mackerel, raw, uncooked	5.27	1.34
Salmon, raw	2.93	0.30
Mackerel, raw, uncooked	4.76	0.41
Pacific saury, raw, uncooked	5.93	0.52
Tuna, bluefin, lean meat, raw	4.10	0.31
Oysters, raw, uncooked	3.87	0.26
Squid, raw, uncooked	2.85	0.57
Shrimp, boiled, dry, whole	16.34	3.89
Shrimp, boiled, dry, w/o shell	19.96	2.05
Fish paste, steamed (Kamaboko)	5.38	0.56
Fish paste, broiled (Kamaboko)	6.70	0.19
Fish paste, broiled (Chikuwa)	3.60	0.46
Fish paste, fried (Satsuma-age)	2.00	0.26
Salted or broiled, dry fish	1.59	1.27
Fishes boiled down in soy sauce	5.64	1.44
Meats		
Beef chuck loin, total edible	8.78	0.58
Chicken thigh, broiler	4.27	0.12
Chicken thigh, broiler, flesh only	3.50	0.21
Swine, inside ham, separable lean	7.50	0.14
Ham, mixed press	6.42	0.36
Sausage, mixed	5.32	1.01

TABLE A-2. Continued.

Food item	Moisture	TDF
	g per 100g edible portion	
Eggs		
Chicken, whole egg, fresh	3.10	0.13
Milk		
Ordinary liquid milk	4.97	0.22
Milk beverage	3.82	0.07
Yogurt, whole milk, unsweetened	3.50	0.10
Lactic acid bacteria beverage	3.50	0.00
Skim milk, domestic	4.92	0.85
Process cheese	4.32	0.58
Vegetables		
Asparagus, green, raw	2.31	1.68
Asparagus, green, boiled, canned	6.52	1.26
Kidney beans, pods, immature, raw	13.77	2.36
Green soybeans, immature, raw	1.27	5.44
Garden pea pods, immature, raw	4.24	2.04
Greenpeas, canned	4.21	7.04
Osaka-shirona, leaves, raw	3.26	1.51
Okra pods, immature, raw	4.07	4.59
Turnip root, raw	12.77	1.30
Turnip root, salted	15.00	1.77
Pumpkin, raw	3.10	2.99
Cauliflower, raw	8.55	1.71
White ground shavings, dry (Kampyo)	7.05	25.84
Chrysanthemums, edible flower, raw	4.30	2.99
Cabbage, raw head	12.00	1.42
Cucumber, whole, raw	14.50	0.85
Nukazuke cucumber pickles, fruit	7.72	1.41
Pot herb mustard, raw leaves (Kyona)	4.48	2.01
Burdock, edible, boiled root	2.80	3.58
Komatsuna, leaves, raw	7.47	1.73
Garland chrysanthemums, leaves, raw	4.11	1.39
Ginger, tuber, raw	11.82	1.83
Sugukina, pickles	11.17	3.78
Water dropwort, leaves, raw	2.54	2.18
Celery, stalk	3.44	1.93
Royal fern, fresh, boiled	6.93	3.45
Japanese radish immature greens	2.80	1.18
Japanese radish greens (Daikon)	4.90	2.74

TABLE A-2. Continued.

Food item	Moisture	TDF
	g per 100g edible portion	
Japanese radish	14.90	1.34
Japanese radish, cut, dry	13.20	17.89
Japanese radish, pickles	11.80	3.79
Broad leaf mustard, leaves	—	2.42
Bamboo shoots, boiled	7.04	2.27
Onions, raw	10.90	1.50
Head lettuce, butter head type	4.87	1.14
Head lettuce, crisp head type	4.10	0.96
Chingen tsuai, leaves, raw	1.69	1.01
Wax gourds, fruit raw	11.70	0.84
Garlic bulbs, sweet pepper, etc., mixed	2.55	8.64
Komatsuna, turnip leaves, perilla	4.48	1.78
Pickled vegetables mix	6.48	3.32
Sweet corn, boiled	2.16	2.01
Tomato, raw	13.30	0.79
Tomato juice, canned	3.67	0.52
Eggplant, raw	8.66	1.66
Rape, flower cluster, raw	—	2.67
Chinese chives, leaves, raw	6.77	1.92
Carrots, root, raw	6.40	2.55
Welsh onion, leaf and leaf sheath	9.29	1.89
Nozawana, pickles, seasoned	8.90	1.69
Chinese cabbage head, raw	14.10	1.09
Chinese cabbage head, salted	2.41	2.03
East Indian lotus root, raw	5.97	1.11
Parsley	2.30	3.14
Sweet pepper, fruit, raw	8.90	1.97
Hirosimana, leaves, raw	6.43	1.34
Japanese butterbur petiole, raw	9.60	0.98
Japanese butterbur petiole, boiled	9.90	1.41
Swiss chard leaves, raw	6.00	2.88
Broccoli, head, raw	2.93	2.68
Spinach, leaves, raw	3.20	2.50
Mytsuba, leafstalk, leaves, green	6.10	2.26
Myoga, bract, flower	5.60	1.41
Brussels sprouts, head, raw	9.87	4.52
Soybean sprouts, raw	2.68	1.56
Mungbean sprouts	5.15	1.20
Blackgram sprouts	11.53	1.39
Lily scale, raw	5.70	6.97
Scallions, sweetened pickles	2.63	8.91
Shallots	5.33	2.17
Braken, fresh, boiled	3.17	3.95

TABLE A-2. Continued.

Food item	Moisture	TDF
	g per 100g edible portion	
Fruits		
Apricot, dried	7.81	8.29
Strawberries, raw	8.60	1.52
Strawberry, jam	1.94	0.76
Figs, raw	13.46	1.46
Satsuma mandarins, common	17.30	1.05
Satsuma mandarin fruit juice	9.12	0.27
Japanese persimmons, raw	9.19	1.60
Japanese persimmons, dried	5.26	10.80
Kiwi fruit, raw	10.03	2.65
Grapefruit, raw w/o membrane	8.50	0.73
Watermelon, raw	5.27	0.22
Japanese plums	19.42	0.77
Japanese pears	3.33	1.07
European pears, w/o skin	5.27	1.74
Pineapple, raw	7.20	0.92
Pineapple, canned	5.02	1.00
Bananas, raw	10.30	1.48
Bananas, dried	6.45	6.47
Grapes, raw	4.40	0.39
Raisins	6.19	4.60
Hybrid melon (Ams), raw	9.50	0.41
Muskmelon, raw	10.70	0.96
Peaches, canned with syrup	9.46	1.47
Apples, raw w/o skin	7.60	1.63
Apple juice, single strength	57.48	0.08
Apple drink	55.53	0.01
Orange marmalade and Apricot jam	4.03	1.38
Fungi		
Enokitake, raw	8.07	2.87
Jew's-ear, black, dried	1.51	74.18
Shiitake, raw, uncooked	8.85	4.54
Shiitake, dry, uncooked	4.40	43.41
Honshimeji, raw	6.69	2.30
Shimeji, raw	5.67	3.09
Nameko, raw, uncooked	6.12	1.80
Common mushroom, raw, uncooked	8.80	1.55
Common mushroom, boiled	6.04	2.23
Hiratake and other fungi, mixed	9.87	3.22

TABLE A-2. Continued.

Food item	Moisture	TDF
	g per 100g edible portion	
Algae		
Green laver, dried	12.40	38.62
Purple laver, dried	4.40	29.68
Makonbu kelp, dried	6.30	28.58
Konbu kelp, salted, dried	5.23	14.61
Agar-agar, dry	16.47	81.29
Hijiki algae, boiled, dried	12.20	54.94
Mozuku algae, raw, desalted	7.03	0.65
Wakame algae, raw	1.30	9.90
Wakame algae, dried	13.10	37.95
Shaved kelp and other algae mixed	3.76	18.34
Beverages		
Coffee, instant	4.16	14.18
Coffee drink, canned	2.87	0.08
Seasonings and Spices		
Consomme, dried	1.89	0.61
Soy-sauce, Koikuti thicker type	8.16	0.81
Worcester sauce, common	6.35	0.39
Worcester sauce, thick type	9.73	1.36
Tomato ketchup	7.37	1.01
Mayonnaise, whole egg type	2.45	0.07
Curry roux	4.26	4.34
Corn soup powder, hash, roux	3.00	4.98
Prepared Foods		
Dip sauce for staked beef	34.94	1.16
Sprinkling mix for rice with tea	1.49	2.03
Miso soup precooked and dried	7.37	6.19
Gyo-za, frozen	7.17	2.10
Croquettes, frozen, potato type	5.04	1.37
Shumai, frozen	2.92	1.33
Hamburger, frozen	6.17	1.22
Meat balls, frozen	8.21	1.59
Processed rice products, mixed	10.52	0.34

Reprinted with permission of T. Nishimune, T. Suminoto, T. Yakusiji, T. Ichikawa, M. Douguchi, S. Nakahara, and N. Kunita. Determination of total dietary fiber in Japanese foods. 1991. *Journal of the Association of Official Analytical Chemists* 74: 350–359.

TABLE A-3. Total Dietary Fiber in Baked Products and Cereals by the AOAC Method.

Food item	Moisture	TDF
	g per 100g edible portion	
Breads, Muffins, and Cakes		
Norwegian flat bread	4.9	16.5
Crispbread, lite rye	7.2	15.9
Matzo, whole wheat	4.8	11.8
Taco shells	7.0	8.1
Melba toast, rye	6.7	7.9
Pita bread, whole wheat	30.0	7.9
Muffins, bran	35.0	7.5
Melba toast, whole wheat	6.1	7.4
Melba toast, plain	5.9	6.8
Muffin, whole wheat English	45.7	6.7
Croutons, plain	6.2	5.2
Croutons, seasoned	4.2	5.0
Matzo, egg onion	8.7	5.0
Hollywoood light bread	37.8	4.8
Boston brown bread	51.3	4.7
Egg bagels	33.0	3.9
Egg bread	30.4	3.8
Croissants, cheese	21.0	3.8
Soft flour tortillas	20.2	3.7
Matzo, plain	7.3	3.5
Coffee cake, crumb topping	29.8	3.3
French toast, frozen	48.1	3.1
Breadstuffing, chicken flavor	60.0	2.9
Gingerbread, mix	38.5	2.9
Croissant, butter	25.1	2.8
Croissant, fruit	45.6	2.5
Coffee cake, fruit	35.6	2.5
Buckwheat pancakes	49.9	2.3
Donuts, yeast raised, glazed	20.3	2.2
Danish, fruit	27.6	1.9
Cornbread, mix	34.4	1.8
Pies		
Pecan pie	23.2	3.5
Apple pie	59.9	2.8
Pumpkin pie	58.2	2.7
Plain cheesecake	45.4	2.1
Chocolate cream pie	47.2	2.0
Cheesecake, no bake	46.2	1.9
Egg custard pie	24.2	1.6

TABLE A-3. Continued.

Food item	Moisture	TDF
	g per 100g edible portion	
Boston cream pie	47.6	1.4
Lemon meringue pie	35.4	1.2
Cherry pie	48.3	0.7
Cookies and Crackers		
Whole wheat crackers (Triscuit)	1.9	10.1
Cookies, fig bars	19.9	4.6
Crackers, graham regular	5.7	3.6
Cookies, chocolate chip	19.6	3.2
Cookies, chocolate sandwich	5.0	3.0
Cookies, butter	3.5	2.4
Cookies, shortbread (Pecan Sandies)	5.5	1.8
Cookies, peanut butter	11.5	1.8
Crackers, honey graham	5.6	1.7
Cookies, Chinese fortune	8.0	1.6
Cookies, vanilla cream	2.9	1.5
Crackers, peanut butter-cheese	3.9	1.1
Miscellaneous		
Bulgur	10.4	18.3
Wheat germ crude	9.5	16.9
Wheat germ toasted	5.8	16.1
Coucous	8.9	9.8
Brown sugar cones	2.9	4.6
Wafer cones	5.6	4.1
Fruitcake	27.0	3.7
Hominy	83.0	2.5
Cornstarch	7.8	0.9
Cereals		
Triticale	11.2	18.1
Barley, scotch	9.2	17.3
Barley, pearled	9.9	15.6
Amaranth	13.3	15.2
Millet	8.2	8.5
Bran		
Corn bran	6.5	82.4
Wheat bran	11.6	42.6
Rice bran	8.2	21.7

TABLE A-3. Continued.

Food item	Moisture	TDF
	g per 100g edible portion	
Oat bran	7.8	17.9
Rice polish	12.0	15.3
Flour		
Dark rye flour	11.1	32.0
Roman Meal	9.1	18.1
Medium rye flour	8.8	14.7
Whole wheat flour	11.5	14.6
Triticale flour	9.4	14.6
Light rye flour	10.6	14.6
Corn flour	10.9	13.4
Amaranth flour	13.0	10.2
Buckwheat flour	12.3	10.0
Masa harina	11.3	9.6
Oat flour	12.2	9.6
Corn meal, degermed	10.3	9.5
All purpose flour (Brand A)	11.2	5.6
All purpose flour (Brand B)	11.3	5.3
Brown rice flour	12.9	4.6
High gluten flour	14.3	4.1
Semolina flour	11.7	3.9
Cake flour	11.6	3.7
Cornmeal	11.0	3.6
Arrowroot flour	11.4	3.4
Rice flour	11.9	2.4
All purpose flour (Brand C)	12.9	2.3
Tapioca flour	11.0	0.9
Rice		
Wild rice	8.4	5.2
Brown rice	11.7	3.9
Instant rice (Brand A)	9.0	2.9
Short-grain rice	11.1	2.8
Glutinous rice	10.1	2.8
Rice-a-roni	71.5	2.5
Instant rice (Brand B)	8.6	2.3
Parboiled rice, dry	10.4	2.2
Medium-grain rice	11.7	1.7
Noodles and Pasta (Uncooked)		
Spaghetti, whole wheat	8.8	10.7

TABLE A-3. Continued.

Food item	Moisture	TDF
	g per 100g edible portion	
Spaghetti, spinach	8.7	10.6
Pasta, artichoke	9.7	9.8
Noodles, spinach-egg	8.5	6.8
Noodles, udon	11.5	5.4
Noodles, chow-mein	0.8	4.7
Macaroni, protein-fortified	8.5	4.3
Pasta, multicolor	10.1	4.3
Noodles, somen	11.3	4.3
Noodles, egg	9.5	4.0
Noodles, Chinese rice	6.2	3.9

Reprinted with permission of *Cereal Foods World* 33:414–418. Cardozo, M.S. and R.R. Eitenmiller. 1988. Total dietary fiber analysis of selected baked and cereal products.

TABLE A-4. Moisture and Dietary Fiber Content of Cooked and Raw Vegetables.

Food item		Moisture	IDF	SDF	TDF
		g per 100g cooked or raw weight			
Cooked					
Artichoke	b	—	3.20	4.36	7.56
Artichoke globe	b	82.5	3.17	4.68	7.85
Asparagus	b	90.5	1.57	0.49	2.06
Aubergine	g	88.3	2.31	1.19	3.50
Bean, snap	b	88.5	2.06	0.83	2.90
Bean, green	b	92.0	2.14	0.71	2.85
Beet, leaves	b	95.1	1.37	0.20	1.57
Beet, roots	b	82.5	2.05	0.55	2.60
Broad bean	s	72.1	4.86	1.08	5.94
Broccoli, tops	b	90.9	2.51	0.56	3.07
Brussels sprouts	b	84.7	4.30	0.75	5.04
Cabbage, savoy	b	94.3	1.77	0.75	2.52
Cabbage, white	b	94.8	2.00	0.71	2.71
Cardoon	b	97.1	1.25	0.29	1.54
Carrots	b	89.0	1.63	1.46	3.08
Cauliflower	b	82.3	1.68	0.70	2.38
Chicory leaves	b	92.4	2.43	1.12	3.55
Eggplant	b	—	5.25	1.32	6.57
Fennel	b	95.4	1.39	0.49	1.87
Jerusalem artichokes	s	79.5	2.02	0.66	2.68
Leek	b	90.3	2.00	0.85	2.85
Mushroom	b	—	3.84	0.33	4.16
Mushroom, field	s	86.0	3.09	0.22	3.31
Mushroom, oyster	s	86.5	4.65	0.34	4.99
Onion, immature	b	90.8	1.20	0.60	1.80
Onion, mature	b	92.3	0.76	0.49	1.25
Pea	s	70.7	5.73	0.59	6.32
Pepper, green	g	87.6	0.93	0.75	1.68
Potato, new	b	80.5	1.05	0.38	1.43
Potato, old	b	78.2	0.85	0.71	1.56
Rape, leaves	b	98.9	1.59	0.61	2.20
Rape, sicilian	b	91.8	2.42	0.84	3.27
Salsola	b	93.3	1.88	0.41	2.29
Spinach	b	93.8	1.64	0.42	2.06
Squash	b	93.2	0.98	0.36	1.34
Stringbean	b	—	2.07	0.86	2.93
Turnip	b	94.7	1.51	0.52	2.03
Raw					
Broad bean		82.9	4.45	0.52	4.97
Cabbage, savoy		88.8	2.53	0.35	2.87

TABLE A-4. Continued.

Food item	Moisture	IDF	SDF	TDF
	g per 100g cooked or raw weight			
Cabbage, white	91.7	2.26	0.33	2.59
Carrots	89.0	2.70	0.41	3.11
Celery	95.7	1.41	0.19	1.59
Cucumber	95.9	0.54	0.21	0.75
Fennel	92.6	1.97	0.25	2.22
Mushroom, field	89.7	2.14	0.11	2.25
Onion, immature	88.8	1.49	0.37	1.86
Onion, mature	92.8	0.88	0.17	1.05
Pepper, green	89.9	1.47	0.41	1.88
Radish	96.6	1.23	0.08	1.30
Tomato	93.8	0.78	0.24	1.02
Turnip	92.9	2.32	0.29	2.61
Salads				
Endive, Belgian	94.7	0.97	0.17	1.14
Endive, curly	94.7	1.40	0.18	1.57
Escarole	93.5	1.86	0.25	2.11
Lettuce, Boston	95.5	1.12	0.18	1.29
Lettuce (Romaine), cos	95.0	1.34	0.13	1.47
"Puntarelle"	94.9	1.49	0.23	1.71
"Radicchio rosso"	90.3	2.47	0.59	2.95

(b)=boiled, (g)=grilled, (s)=sauteed

Reprinted with permission of C. Lintas and M. Cappeloni. 1988. Content and composition of dietary fibre in raw and cooked vegetables. *Food Sciences and Nutrition* 42F:117-124, and E.D. Toma, A. Clementi, M. Marcelli, and C. Lintas. 1988. Food fiber choices for diabetic diets. *American Journal of Clinical Nutrition* 47:243-246.

TABLE A-5. Total Dietary Fiber Content of Fruit, Bread, Grains, Cereals, Rices, Legumes, and Vegetables by the AOAC Method.

Food item	TDF
	g per 100g fresh edible portion
California seedless raisins[a]	6.80
Japanese plum:	
Black Doris	1.50
George Wilson	1.44
Strawberry, tioga	1.82
Blackberry, thornless	5.65
Blackcurrant, magnus	5.61
Raspberry, marcy	3.06
Boysenberry, thorny strain	4.27
Blueberry, jersey	3.41
Tamarillo, yellow	3.26
red	3.07
Feijoa, triumph	3.01
mammoth	2.84
Kiwifruit, bruno	2.31
hayward	2.63
Babaco, standard	0.74
Rockmelon, winstones hybrid 1	0.58
Pear, william bon chretien	2.25
packham's triumph	3.32
Apples	
Granny smith	1.91
Sturmer pippin	2.12
Red delicious	1.78
Golden delicious	1.90
Gala	1.51
Cox's orange pippin	1.88
Delicious	1.67
Stonefruit	
Peach, red haven	1.86
flamecrest	1.68
springcrest	1.65
Nectarines, red gold	1.64
fantasia	1.60
Apricot, moorpark	2.17
steven's favourite	1.98
sundrop	1.88
Cherry, dawson	1.30
bing	1.21

TABLE A-5. Continued.

Food item	TDF
	g per 100g fresh edible portion
Bread	
White bread[b]	2.60
Wholemeal bread[b]	5.70
75% bulgar: 25% wheat flour[c]	4.71
50% bulgar: 50% wheat flour[c]	4.64
100% rye[c]	5.79
Pumpernickel bread (Kasseler)[b]	10.05
Pumpernickel bread (Dimpflmeir)[b]	7.40
Rye bread[b]	21.10
Grain and Cereal Products (Uncooked Weight)	
Spaghetti[b]	4.80
Macaroni[b]	4.80
Pearl barley[b]	13.50
Bulgur wheat[b]	9.10
AACC wheat bran[d]	44.10
Breakfast Cereals (Uncooked)	
Oat bran[d]	17.90
Oat bran[b]	16.40
Rice chex[b]	0.01
Corn chex[b]	3.10
Low fiber wheat cereal[e]	0.88
Bran flakes[e]	18.60
All-bran[e]	31.60
Rices	
Parboiled rice[b]	2.20
Long-grain rice[b]	2.00
Instant rice[b]	1.90
Legumes (Canned Weight)	
Red lentils[b]	11.10
Kidney beans[b]	4.60
Romano beans[b]	5.20
Chick peas[b]	4.60
Red kidney beans (Phaseolus vulgaris)[d]	19.60
Ground whole soybean[f]	17.26

TABLE A-5. Continued.

Food item	TDF
	g per 100g fresh edible portion
Whole soybean extrudate*[f]	16.85
Whole soybean press cake*	19.35
*from ZYB-78 oil expeller	
Vegetables	
Instant potato[b]	7.90
Pea, puget	5.13
Pea, victory freezer	4.84
Broadbean, sevill variety	5.86
Greenbean, gallatin 50	3.04
Mushroom, a. bisporous	1.36
Sweetcorn, reliance	1.94
Onion, pukakoha longkeeper	1.67
Greenbean, greenpak	2.78
Tomato, UC-82	1.05
Asparagus, mary washington	1.88
Carrot, chantenay red-cored	2.54

Unlettered values are from the paper of F.R. Visser and C. Gurnsey. 1986. Inconsistent differences between neutral detergent fiber and total dietary fiber values of fruits and vegetables. *Journal of the Association of Official Analytical Chemists* 69:565–577.

[a]Payne, T.J. 1987. The role of raisins in high-fiber Meusli-style formulations. *Cereal Foods World* 32:545–547.

[b]Jenkins, D.J.A., D. Cuff, T.M.S. Wolever, D. Knowland, L. Thompson, Z. Cohen, and E. Prokipchuk. 1987. Digestibility of carbohydrate food in an ileostomate: Relationship to dietary fiber, in vitro digestibility, and glycemic response. *The American Journal of Gastroenterology* 82:709–717.

[c]Jenkins, D.J.A., R.D. Peterson, M.J. Thorne, and P.W. Ferguson. 1987. Wheat fiber and laxation: dose response and equilibration time. *Gastroenterology* 82:1259–1263.

[d]McBurney, M.I. and L.U. Thompson. 1987. Effect of human faecal inoculum on in vitro fermentation variables. *British Journal of Nutrition* 58:233–243.

[e]McBurney, M.I., L.U. Thompson, and D.J.A. Jenkins. 1987. Colonic fermentation of some breads and its implication for energy availibility in man. *Nutrition Research* 7:1229–1241.

[f]Nelson, A.I., W. B. Wijeratne, S.W. Yeh, T.M. Wei, and L.S. Wei. 1987. Dry extrusion as an aid to mechanical expelling of oil from soybeans. *Journal of the American Oil Chemists Society* 64:1341–1347.

TABLE A-6. Dietary Fiber Content of Miscellaneous Cereal Grain Food Products.

Product	Moisture	IDF	SDF	TDF
	g per 100g of product			
Bread[abc]				
White bread	36.7	—	1.00	3.00
Pumpernickel	40.0	—	1.79	7.14
White pita	26.5	—	1.54	2.70
French	38.0	—	1.03	3.09
Sourdough	36.4	—	1.01	3.01
Raisin	30.1	—	0.96	4.27
White "lite"	44.3	—	0.49	12.76
Mixed grain "lite"	39.3	—	0.91	13.56
Corn	29.0	—	0.48	2.48
Cellulose containing	41.5	—	1.10	9.90
Hamburger buns	35.0	—	0.89	2.59
French rolls	30.8	—	1.01	3.23
Bagels (plain)	34.5	—	0.94	2.47
Biscuits (refrigerated)	29.7	—	1.10	1.77
Bread sticks	4.6	—	1.24	3.04
Corn tortillas	43.8	—	0.95	4.33
Flour tortillas	29.3	—	1.02	2.30
Croissants	20.4	—	0.86	2.32
Oatmeal	36.3	—	1.20	3.70
Tortillas, corn	43.8	—	0.90	4.30
Rye, German	36.8	—	1.50	8.30
Cracked wheat	35.2	—	0.90	6.70
Bran containing	39.1	—	0.80	5.40
Whole wheat 100%	39.7	—	1.10	8.10
Mixed grain	40.3	—	0.60	5.60
Multigrain	39.5	—	1.30	9.60
Multigrain + DPV-1[1]	38.3	—	1.40	3.40
Multigrain + DPV-2	38.8	—	1.20	4.90
Crackers and Snacks[cd]				
Soda crackers made with pastry flour	4.0	1.30	1.40	2.70
Soda crackers made with bread flour	4.0	2.00	1.70	3.70
Saltines	3.0	—	1.21	2.34
Honey grahams	3.0	—	1.27	2.98
Whole wheat crackers	1.9	—	1.77	10.86

[1]DPV is dried powdered vegetables (DPV-1 is light; DPV-2 is dark). Vegetables include carrots, pumpkins, lettuce, artichoke, celery, cabbage, parsley, cauliflower, and kelp.

TABLE A-6. Continued.

Product	Moisture	IDF	SDF	TDF
	g per 100g of product			
Snack crackers[2]	3.3	—	0.97	2.04
Cheese crackers	3.5	—	1.08	2.45
Wheat crackers	1.7	—	1.13	4.02
Melba toast (white)	5.1	—	1.53	6.15
Melba toast (wheat)	5.9	—	1.79	8.87
Corn chips	1.1	—	0.43	4.29
Pretzels (hard)	3.5	—	0.94	3.66
Pretzels (soft)	32.7	—	0.93	2.49
Taco shells	4.2	—	0.91	6.33
Breakfast Items[c]				
English muffins	13.4	—	1.28	4.23
Blueberry muffins	33.6	—	0.53	1.66
Bran muffins	23.3	—	0.75	3.25
Oatbran muffins	26.3	—	0.54	1.69
Waffles (frozen)	37.7	—	0.69	1.89
Puffed wheat	6.9	—	2.38	7.53
Toasted oats	3.5	—	2.76	7.02
Raisin bran	7.6	—	2.37	13.50
All bran + fiber	3.1	—	2.58	51.20
Shredded wheat	7.1	—	1.55	12.53
Cream of wheat	9.5	—	1.57	3.82
Sweet Goods[c]				
Oatmeal cookies	4.9	—	1.10	2.57
Chocolate chip cookies	4.1	—	0.70	2.58
Vanilla wafers	5.5	—	0.81	1.50
Ginger snaps	2.9	—	1.02	1.54
Cream-filled chocolate cookies	1.6	—	0.97	3.48
Shortbread cookies	3.8	—	0.86	1.79
Brownies	12.8	—	0.55	2.54
Angel food cake	35.5	—	0.28	0.77
Devil's food cake	30.4	—	0.74	2.48
Ice cream cones	6.0	—	1.32	3.02
Apple pie	44.9	—	0.67	1.57
Cream-filled cakes[3]	22.1	—	0.31	0.84
Cinnamon rolls	19.2	—	0.75	2.41
Cake doughnuts	24.5	—	0.57	1.70
Yeast-raised doughnuts	21.1	—	0.52	1.21

[2]with sprayed on fat
[3]nondiary cream filling

TABLE A-6. Continued.

Product	Moisture	IDF	SDF	TDF
	g per 100g of product			
Pasta and Rice[c]				
Spaghetti (dry)	9.2	—	1.48	3.64
Spaghetti (cooked)	73.4	—	0.40	1.30
Egg noodles (dry)	9.3	—	1.39	3.91
Egg noodles (cooked)	67.3	—	0.53	1.83
White rice (dry)	8.7	—	0.27	1.25
White rice (cooked)	77.5	—	0.03	0.71
Wheat[ce]				
Hill 81	—	—	—	10.50
DDGS[5]	—	—	—	34.60
Tyee	—	—	—	11.60
DDGS	—	—	—	33.90
Red wheat blend	—	—	—	9.90
DDGS	—	—	—	34.80
Pastry flour	—	—	—	1.60
DDGS	—	—	—	14.40
Straight flour	15.0	—	1.09	2.51
Patent flour	14.7	—	1.31	2.50
Whole wheat flour	15.9	—	1.32	10.24
Germ	14.3	—	1.14	9.31
Bran	15.6	—	2.10	44.03
Wheat Flour (80% Extraction)[f]				
Native	—	—	—	4.60
Extruded	—	—	—	4.50
Drum-dried	—	—	—	4.70
Wheat Flour (98% Extraction)[f]				
Native	—	—	—	11.70
Extruded	—	—	—	11.60
Drum-dried	—	—	—	10.60
Wheat bran[g]	—	—	—	38.00
Bakers' patent wheat flour[h]	—	—	—	3.40
Wheat bran	—	—	—	54.20
Corn				
DDGS[5]	—	—	—	32.00

TABLE A-6. Continued.

Product	Moisture	IDF	SDF	TDF
	g per 100g of product			
Corn bran	—	—	—	90.30
Corn fiber[i]	—	—	—	61.70
Cotton seed hulls[i]	—	—	—	73.70
Oat hulls[i]	—	—	—	72.20
Soybean hulls[i]	—	—	—	79.60
Field pea hulls[f]	—	—	—	82.30
Barley	—	—	—	21.70
DDGS[5]	—	—	—	84.70
Brown sorghum	—	—	—	10.10
DDGS[5]	—	—	—	67.80
Oat bran[g]	—	—	—	26.40
Wild oat bran[h]	—	—	—	20.00

[5]DDGS = Distillers' dried grain with solubles.

[a]Ranhotra, G., J. Gelroth, and P. Bright. 1987. Effect of the source of fiber in bread-based diets on blood and liver lipids in rats. *Journal of Food Science* 52:1420–1422.

[b]Ranhotra, G.S., J.A. Gelroth, and P.H. Bright. 1988. Effect of the source of fiber in bread on intestinal responses and nutrient digestibilities. *Cereal Chemistry* 65:90–12.

[c]Ranhotra, G., and J. Anderson. 1989. Soluble fiber in bakery products lowers blood cholesterol in men. *Preliminary Report, American Institute of Baking* pp. 1–6.

[d]Ranhotra, G., and J. Gelroth. 1988. Soluble and insoluble fiber in soda crackers. *Cereal Chemistry* 65:159–160.

[e]San Buenaventura, M.L., F.M. Dong, and B.A. Rasco. 1987. The total dietary fiber content of wheat, corn, barley, sorghum, and distillers' dried grains with solubles. *Cereal Chemistry* 64:135–136.

[f]Schweizer, T.F., and S. Reimann. 1986. Influence of drum-drying and twin-screw extrusion cooking on wheat carbohydrates. I. A comparison between wheat starch and flours of different extraction. *Journal of Cereal Science* 4:193–203.

[g]Chen, H., G.L. Rubenthaler, H.K. Leung, and J.D. Baranowski. 1988. Chemical, physical and baking properties of apple fiber compared with wheat and oat bran. *Cereal Chemistry* 65:244–247.

[h]Sosulski, F.W., and K.K. Wu. 1988. High fiber breads containing field pea hulls, wheat, corn and wild oat brans. *Cereal Chemistry* 65:185–191.

[i]Hsu, J.T., D.B. Faulkner, K.A. Garleb, R.A. Barclay, and L.L. Berger. 1987. Evaluation of corn fiber, cotton seed hulls, oat hulls and soybean hulls as roughage sources for ruminants. *Journal of Animal Science* 65:244–255.

Index